A South Downs National Park

Public consultation report

Distributed by:
Countryside Agency Publications
PO Box 125
Wetherby
West Yorkshire LS23 7EP

Telephone   0870 120 6466
Fax   0870 120 6467
Minicom   0870 120 7405 (for the hard of hearing)
www.countryside.gov.uk

Contents

Part B: The draft boundary

Foreword

The South Downs is a very special place, much loved for its natural beauty, diversity and tranquillity. Yet national, regional and local pressures threaten to diminish the very qualities that make the area special.

The Countryside Agency believes that national park designation is the best way to protect the Downs, build on the achievements of the past, and conserve the area for future generations. The legal status of a national park will give this valuable area the highest degree of protection and will bring the management arrangements and resources it needs.

A South Downs National Park would be different from the existing national parks. It would have more visitors and more people living within its boundary, more intensive agriculture and a greater need for landscape restoration, more development pressures and more planning applications than any other. In short, the area is under more pressure.

Yet none of the challenges these pressures bring are unique to the South Downs, other national parks are also dealing with them to a greater or lesser degree. There is nothing in the South Downs that a national park authority - with an appropriate level of resources, special legal powers, membership, responsibilities and purposes - could not address.

Our vision is of a South Downs National Park working within a different social, economic and rural environment, distinct from those in the more remote upland national parks. We are looking at the Downs with a fresh eye and working with a new concept: a modern interpretation of national parks in the 21st century. This means that national parks need no longer be predominantly in the rugged upland countryside of England and Wales, but can bring benefits equally well in other outstanding landscapes.

We have already done a great deal of work towards making this vision a reality, including discussions with key bodies, working groups and seminars. This has been invaluable in shaping our thinking and I thank everyone who has contributed so far. This document draws on that work and sets out our preferred options for the administrative arrangements for a South Downs National Park Authority, and presents our first draft of the national park boundary. However, this is not our final view.

We now want to hear from everyone who has an interest in the South Downs and cares about its future. I invite you to send us your comments (see Chapter 1 for how to do so). This document is widely available within the South Downs and is being sent to key bodies with an interest, locally and nationally. We have also published a consultation booklet, and are holding public events and other meetings throughout the consultation period.

All the documents and details about the events are on our website:

www.countryside.gov.uk/proposednationalparks

I look forward to hearing your views on the best way to set up a national park authority that is right for the South Downs, and about the draft boundary we have proposed.

Ewen Cameron
Chairman

1 Introduction

The challenge is to create a national park, and a national park authority, that is right for the South Downs in the 21st century.

The Countryside Agency has two main tasks:

- identifying the statutory boundary for a South Downs National Park and making a South Downs National Park Designation Order;
- advising the Secretary of State for Environment, Food and Rural Affairs on administrative arrangements for a national park authority that reflect the South Downs' special circumstances, for his/her consideration when making the South Downs National Park Authority Establishment Order.

This consultation document invites your views on both of these tasks.

A South Downs National Park

The Countryside Agency has a statutory responsibility under the National Parks and Access to the Countryside Act 1949 for designating national parks (see Appendix A for national park legislation).

In April 2000 the Agency decided that it should begin the process of designating a national park in the South Downs because of its great natural beauty, national significance for outdoor recreation and the need for management by a special authority. The South Downs is a special place, and one that merits the highest legal protection and the best possible management for the future.

The Agency believes that designation as a national park is the best way to protect and conserve one of England's finest landscapes for present and future generations, building on the achievements of the past.

National park status will bring to the South Downs the best measure for countryside protection available in the UK, providing the greatest powers for action, the most resources and the highest standards of conservation and management, to enhance the Downs and help people to enjoy its special qualities.

The Countryside Agency now has two main tasks. It must identify the right boundary for a South Downs National Park, drawing in land that meets the criteria defined by law and which will benefit from being included within a national park. We also need to ensure that a national park authority is equipped with the right tools to address the particular management challenges of the South Downs.

We need to create the right type of national park for this area, one that is able to address its special needs, and one that is right for the 21st century.

We would like to hear your views

Administrative arrangements

National park status will give the South Downs the best possible protection and management because there will be a focussed body - a national park authority - whose purpose will be to care for the area and to help people enjoy it. The authority will have special powers and the resources from government to achieve this.

The Countryside Agency has carried out work to determine the particular administrative issues that will be important to a South Downs National Park Authority, and to identify solutions for addressing these. This process involved the help of a number of technical advisory and topic groups, established to give factual and technical advice to the Agency, as well as seminars and meetings with local and national partners (see Appendix D).

We invite your views on how a South Downs National Park Authority should be set up.

After considering your responses, the Countryside Agency will draw up draft recommendations for the Secretary of State on the best administrative arrangements for a South Downs National Park Authority. We will invite local authorities to comment on these, and then after considering their responses will submit our proposals to the Secretary of State, along with the Designation Order that creates the national park. The Secretary of State will then decide how to set up the national park authority, and also what particular guidance it should be offered about how to carry out its duties.

This consultation is not concerned with the policies and actions that a national park authority may carry out. Policy matters will be determined by the authority itself when established. As it will consult widely on these, they are not addressed in detail in this document. The Countryside Agency will have a role in advising on the development of such policies, as part of its statutory responsibility for national parks generally. Information and advice gathered from the technical advisory and topic groups will also be passed on to a future national park authority.

We would like your views on the proposed administrative arrangements

The Countryside Agency is seeking views on the administrative issues set out in Part A, Chapter 4. We would also like to hear your views if you think there are any other issues that ought to be considered. We would like to know whether you consider that our preferred approach to addressing these issues is the most appropriate to reflect the special circumstances of the South Downs. If not, is one of the other options better, or are there others that we should consider?

In your responses, please state fully the reasons for your view, and explain how you believe your preferred option will ensure that a South Downs National Park Authority can best carry out its functions. It is important that you take into account the prime purposes of a national park to conserve and enhance the countryside and to help people enjoy it, and the special circumstances of the South Downs. These are explained more fully in Chapters 2 and 3.

To help you structure your response, for each issue we ask some specific questions. For ease of reference, we have listed all of the issues and questions in Chapter 5. To help us analyse responses, please indicate to which questions each of your responses refers (eg 2A).

Defining the national park boundary

The Countryside Agency's other main task is to define the area of the national park. This process has been carried out in two stages.

The first stage was to identify the different types of landscapes that make up the South Downs and to assess each one against the national park criteria. Those areas that met the criteria formed an 'area of search' for the boundary. This area of search was agreed in March 2001.

The next stage was to identify a precise draft boundary, around the edge of the area of search, using field-by-field analysis.

Throughout this process, the statutory national park criteria, the Agency's policy and approach to identifying national park boundaries (see Chapters 6 and 7) were applied when deciding which areas to include or exclude.

This document sets out in full our draft national park boundary, and invites you to comment on it. National park boundaries need to be based on the statutory criteria set out in legislation - natural beauty and opportunities for open-air recreation. The document describes these and the other factors we took into account in drawing the draft boundary. It explains why we have included certain areas and not included others.

We have divided the boundary into the 49 sections, listed opposite, to make it easier for you to respond and for us to consider your comments. There is a key map which shows all 49 sections of the boundary on page 93. There are also more detailed maps at the back, showing each section in detail.

The boundary identified in this document is a draft.

We now want to hear from everybody (individuals, local authorities and other organisations) with a view, whether, **on the basis of national park criteria and on the Agency's policy and approach to defining national park boundaries set out in this document** (see Part B), they think the draft boundary is right and, if not, how it should be changed and why. We will look very carefully at the **reasons** people give for their views.

We are consulting widely to ensure that the final boundary is the best for the future of the South Downs. We will consider very carefully all of the comments we receive. We will then publish a proposed boundary and will consult the local authorities formally (as the legislation requires) in 2002.

We will consider local authority responses and decide whether any further changes are needed before preparing a National Park Designation Order (under the National Parks and Access to the Countryside Act 1949) to submit to the Secretary of State in summer 2002.

We would like your views on the draft boundary

We specifically need to know whether you agree with the draft boundary we have identified or whether you think it should be changed. In either case, please give your reasons.

You may want to comment on more than one section. Please make it clear to which section(s) your comments apply by using the relevant boundary section letter(s), eg. section A, map 2. You may also want to comment on whether or not broad areas (for example the Low Weald) should be included within the Park, as well as whether the draft boundary line is in the right place.

In all cases, whether you agree with our draft boundary or wish to propose a change, it is important that you explain your reasons. Please set these out as clearly as possible. It is very important that your reasons relate to:

a) the statutory criteria for the designation of national parks, which are:

- **their natural beauty;**
- **the opportunities they afford for open-air recreation; and**

b) the Countryside Agency's policy and approach to defining national park boundaries set out in Part B of this document.

We will not be able to take responses fully into account unless they relate to these points. Chapter 6 shows how we interpreted the criteria in arriving at the draft boundary.

You may wish to comment particularly on these points, which are based on the Agency's policy on applying the criteria and our approach to identifying a boundary on the ground – see Table 1 (page 38) and Chapter 8.

If you wish to propose an amendment to any part of the boundary, please describe the alternative line you suggest in sufficient detail so that it can be marked on a map. A copy of a map showing your suggestions would be ideal.

Finding out more

During the consultation period we will be meeting with key partners and groups with a particular interest in the Downs, and will be holding a number of roadshows that are open to everybody. These roadshows will provide opportunities to discuss the proposals and ask questions of Countryside Agency staff before making a response to the consultation.

Details of these events and further information will be made available in the local press, from local authorities (including parish councils), and also on the Countryside Agency's website:
www.countryside.gov.uk/ proposednationalparks

We have also produced a public consultation booklet 'A South Downs National Park: we would like to hear your views' (CA 90) that will be made widely available throughout the area and beyond. This booklet asks the same questions as appear

in this publication and includes a comment form (CA 90/F) for responses that you may wish to use. Copies of both the publication and the comment form are available from:

Countryside Agency Publications,
PO Box 125, Wetherby,
West Yorkshire LS23 7EP.
Tel: 0870 120 6466;
Fax: 0870 120 6467;
Minicom: 0870 120 7405.

The booklet is also available on the Countryside Agency's website.

Where and when to send your comments

Please send your comments on the proposed administrative arrangements and/or the draft boundary to:

A South Downs National Park - Public Consultation,
PO Box 33299,
London SW1H 0WF

or by email via the Countryside Agency's website:
www.countryside.gov.uk/ proposednationalparks
as soon as possible, but to arrive not later than **28 February 2002.**

Please note that your response will be retained by the Countryside Agency. Unless you request otherwise, it may be made available for other people to see.

2 National Parks in England and Wales

National parks are not unique to Britain, but are an internationally accepted concept designed to protect a nation's finest landscapes and natural heritage and enable people to enjoy and understand them. In Britain this includes safeguarding the cultural landscape, a countryside where people live and work and whose way of life over many generations shapes the distinctive appearance of that landscape and its settlements.

National parks in England and Wales are designated under the National Parks and Access to the Countryside Act 1949, as amended by the Countryside Act 1968 and the Environment Act 1995. They form part of an internationally recognised network of protected landscapes with six management categories defined by the IUCN (the World Conservation Union). Like other English and Welsh national parks, the South Downs will be classified by the IUCN as a Category V Protected Landscape -

"an area managed mainly for landscape conservation and recreation, but recognising past and present human use".

The management objectives of Category V landscapes can be summarised as:

"to maintain significant areas which are characteristic of the harmonious interaction of nature and culture, whilst providing opportunities for public enjoyment through recreation and tourism, and supporting the normal lifestyles and economic activity of these areas. These areas also serve scientific and educational purposes, as well as maintaining biological and cultural diversity".

There are currently eight national parks in England (Dartmoor, Exmoor, The Lake District, Northumberland, The North York Moors, The Peak District, The Yorkshire Dales and The Broads[1]) and three in Wales (Brecon Beacons, Pembrokeshire Coast, and Snowdonia). As well as the South Downs, the

Countryside Agency is taking forward designation in the New Forest. Two Scottish national parks, the Cairngorms, and Loch Lomond and The Trossachs, are in the process of designation.

During the 50 years since designation of the first English national park[2], they have protected landscapes from inappropriate development, whilst helping people to enjoy and understand their special qualities. National parks have also helped to sustain the farming practices that shaped them and to retain rural communities and a living, working countryside.

Innovative approaches to management have been pioneered in our national parks and they have become exemplars in conservation, recreation management and planning protection. Many of these new approaches have now been applied outside the national parks' own boundaries, informing the management of all of our countryside.

---

1 Although the Broads are not designated under the 1949 national park legislation, they have equal status under the Norfolk and Suffolk Broads Act 1988 and are considered to be a member of the national park family.
2 The Peak District in 1951.

National park criteria and their application

Section 5 of the 1949 Act defines national parks as "extensive tracts of country in England and Wales" which are to be designated because of:

"(a) their **natural beauty;** and (b) the **opportunities they afford for open-air recreation,** having regard both to their character and to their position in relation to centres of population".

The first seven English national parks were designated under this legislation, which fixed their dual purposes of conservation and recreation and led to the family of national parks we are familiar with today. They are outstanding landscapes, managed primarily to conserve their natural beauty (which includes their wildlife, history and cultural heritage, as well as the way the countryside looks), and made accessible in ways that are compatible with conservation and which do not harm their special qualities.

In February 2000 the Countryside Agency Board agreed a new policy for applying national park designation criteria[3], to make sure that the parks remain up to date and continue to reflect modern requirements. It agreed that the key issues to be considered in designating new national parks (other than natural beauty) were the potential of an area to provide a markedly superior recreational experience and whether it would benefit from the special management

that designation would bring, over and above existing arrangements.

The decision, and the thinking behind it, opened the horizons for national parks, and suggested that they could equally bring benefits to less rugged, more populated and more managed lowland landscapes. An important new consideration, then, is the extent to which an area meets people's current countryside requirements by providing quality, open-air recreational experiences, close to large centres of population. The Countryside Agency believes that the South Downs has these attributes.

The role of a national park authority

A national park authority's primary purposes (as defined by the Environment Act 1995) are:

- to conserve and enhance the natural beauty, wildlife and cultural heritage of the national park;
- to promote opportunities for the public to understand and enjoy the national park.

In pursuing these purposes, a national park authority has two legal duties:

- to foster the economic and social well-being of the communities within the national park;
- to give greater weight to the first purpose, should there appear to be any conflict between the two.

Creation of a South Downs National Park Authority would bring an independent body with clearly focused statutory purposes and resources for conservation and recreation, with a remit to manage the whole Downs area in an integrated way.

A national park authority:

- is permanent, so can develop long-term strategies and build up long-lasting partnerships to achieve real results;
- is focused - the two purposes of national parks provide a unique focus that results in positive action;
- brings significant new national funding and expertise, as well as national and international status;
- can provide overall recreation management, taking a holistic view of recreation and visitor patterns;
- initiates projects, and acts directly itself to achieve national park purposes;
- acts as a catalyst, bringing partners together and attracting external funding for projects.

---

3 Countryside Agency, National park designation: A review of how the criteria are applied, Board paper, February 2000 (AP 00/03).

What could a South Downs National Park Authority do?

A national park authority can act to:

* address a wide range of conservation and visitor management issues across the whole of the national park;
* conserve the natural environment and cultural heritage;
* manage recreational use of the area by visitors and local people in line with national park purposes;
* promote understanding through information, interpretation and education;
* provide a countryside management service, including rangers and volunteers, and land management;
* initiate projects, act directly or work with others to encourage joint action;
* promote good practice;
* involve local people;
* advise and fund others to achieve national park purposes;
* become a test bed for new approaches to land and visitor management.

Other national park authority functions include:

* forward planning, through structure plans, local plans and national park management plans;
* development control;
* acting as an advocate for the park at an international, national and regional level;
* providing training for members and staff development.

What do the existing national park authorities do?

For examples of actual work undertaken by national parks, visit the website of the Association of National Park Authorities (ANPA), or those of individual national parks. National parks also produce annual reports, newsletters and other publications about their work.

www.anpa.gov.uk

www.breconbeacons.org

www.broads-authority.gov.uk

www.dartmoor-npa.gov.uk

www.exmoor-nationalpark.gov.uk

www.lake-district.gov.uk

www.northumberland.org

www.northyorkmoors-npa.gov.uk

www.peakdistrict.org

www.pembrokeshirecoast.org

www.snowdonia.org

www.yorkshiredales.org.uk

National park authorities are regarded as special purpose local authorities. They are the planning authority for their area. In other respects they do not replace the work of the local authorities, which continue to have the prime responsibility for a wide range of matters, including education, transport, social services, housing, waste collection and economic development. To deliver their shared objectives, national park authorities and local authorities work closely together on, for example, transport plans and community strategies.

There is wide recognition in national parks that issues of environmental conservation, recreation and economic development in the countryside are closely intertwined. A healthy rural economy is a prerequisite of a well-maintained countryside, while sustainable forms of tourism that respect an area's environmental capacity can bring important income to support rural jobs and services. So, in carrying out its conservation and recreation role, a national park authority would also be involved in working with local people and businesses.

Tourism provides many more jobs than traditional employment in farming and forestry, however it is this basic husbandry of the landscape that encourages visitors. In turn, visitors spend money that sustains local services like shops, pubs and public transport, which safeguard and create jobs. National park authorities work proactively with local authorities and government agencies to develop locally affordable housing and other rural services.

A South Downs National Park Authority, through ANPA, would be a member of a national and international family. ANPA brings together the national park authorities, representing them at national and international level and assisting their dealings with government and other statutory agencies. ANPA helps national park authorities to share best practice, and works to increase public understanding of the purposes of national parks.

National park management plans

National park authorities are required to produce a national park management plan, in wide consultation with local people and organisations. The plan looks at the needs of the area as a whole and seeks ways of balancing conflicting interests within an overall requirement to conserve and enhance the area's natural beauty, wildlife and cultural heritage.

This management plan is for the national park as a whole, and not just for work undertaken by the national park authority, so it is prepared in consultation with other bodies that also play a part in its delivery.

Resources

All of the activities that a South Downs National Park Authority would carry out, notably good visitor management, positive conservation work, and education and interpretation, cost money.

The annual budget of a South Downs National Park Authority will be decided when it is created and its responsibilities have been determined. Government funds for national park authorities are determined by a formula and depend upon the issues the national park needs to address and the work it will carry out.

Seventy-five per cent of the costs of the South Downs National Park would come directly to a national park authority from central government. The remaining twenty-five per cent would come via the local authorities in a South Downs National Park that appoint members to the national park authority. However, the local authorities' contributions would be covered by their annual central government grant, so in practice there are no additional costs locally.

National parks can also raise income from their own trading activities, for example by selling literature, and can attract grants from a wide range of other sources (including the European Union and Lottery funds). National parks can often attract up to a third of their budget this way, on top of the direct funds from government.

3 The South Downs

The South Downs extends over 116 km from Winchester to Eastbourne. Regarded by many as one of England's finest landscapes, the Downs embrace a variety of contrasting landscapes, most notably the rolling chalk downland, and chalk escarpment, with precipitous coombes and steep-sided valleys, together with a dramatic coastline. The area also includes ancient hanging woodlands, wild heathlands, secretive Wealden ghylls and unspoilt river valleys and wetlands.

This diverse, quintessentially English lowland landscape, the area's unique features, and the high degree of accessibility that allows a large number of people to enjoy the landscape, are at the heart of the special appeal of the South Downs.

The value of the South Downs as a national resource has long been recognised, with people seeking national protection and status for the area for over 100 years (see Appendix B). The South Downs were identified as a possible national park in 1947. In the 1960s, both the East Hampshire and Sussex Downs were designated as Areas of Outstanding Natural Beauty (AONBs), and the debate has continued into the 1980s and 1990s about the merits of designating a South Downs National Park.

This dramatic landscape has a distinct image (open downland, extensive views, heaths, dramatic coast) that is recognised locally, nationally and internationally. Indeed, the South Downs were chosen to represent the image of England fought for in the Second World War.

This highly valued landscape is especially treasured for its beauty, cultural associations and recreational potential by those living in and near to the Downs. The Downs provide local countryside for over one million people in the towns and cities close by, including Winchester, Brighton, Petersfield and Chichester. And for some 10 million people who live within an hour's journey time of the Downs.

In contrast with many of the existing national parks, the South Downs are in a very busy, highly pressured part of the country. The pace of life is fast and growing, and people who live and work in the South East need the chance to get away from it all. So a new national park here would not only be a place of great beauty, but would also continue to provide opportunities for a wide range of leisure pursuits, and a countryside experience not otherwise available in the region.

Pressures the area faces
The South Downs is an area of outstanding countryside within the busy and prosperous South East. This leads to many pressures, and to the need for a balance to be struck between conservation and sustainable economic growth.

Development
The demands of a growing modern economy mean that the area is under increasing pressure for development - of roads, industry and other infrastructure, as well as housing.

Managing the South Downs today

The South Downs today as AONBs are cared for by the Sussex Downs Conservation Board and the East Hampshire Joint Advisory Committee. Their work, supported by the Countryside Agency and local authorities, will continue during the designation process.

These two groups work together and in partnership with other organisations to protect, conserve and enhance the natural beauty of the South Downs, encourage quiet enjoyment, and promote sustainable economic and social development. They are funded by the local authorities, and by grants from the Countryside Agency and other bodies.

The creation of these two groups in the early 1990s - in particular the Sussex Downs Conservation Board, which was an experiment in managing AONBs - was in recognition of the special management needs of the Downs.

A national park authority would build on their work, continuing much of what they already do. But it would benefit from greater powers, resources and permanency, and so would be better able to tackle the long-term challenges facing the Downs.

- The national park would be located within the heart of the European Zone for economic development.
- The greatest development pressure in the South East is the demand for housing.
- There would be more planning applications than in other national parks. The South Downs is likely to include more towns and urban areas within and close to its boundary.

To see the role that the Agency has suggested that a South Downs National Park Authority could play in forward planning and development control, see page 19.

Landscape, wildlife and cultural heritage

The South Downs area has a higher proportion of intensively managed and cultivated farmland than in any of the existing national parks. There are about 80,000 hectares of arable land within the two AONBs[4]. This is greater than the total amount of agricultural land (arable and pasture) in any other national park. The Lake District National Park has about 76,000 hectares of land in agricultural use[5].

Farming has shaped this lowland landscape, yet recession is now hitting the industry. The amount farmers receive for their produce has fallen sharply, as have the subsidies that they receive. The weather, BSE and Foot and Mouth Disease have all had, and are continuing to have, a devastating effect on agriculture. Such changes will undoubtedly impact upon how the Downs might look in the future.

With the loss of traditional management techniques over the last 50 years the landscape has suffered considerable wildlife and habitat loss. The remaining biologically important habitats are small, fragmented, frequently under-managed and at risk. Agri-environment schemes have helped to halt this decline, but there has been limited restoration. While 6,000 hectares of arable land has been reverted back to grassland under the Environmentally Sensitive Area scheme, only about 10% (856 hectares)[6] has been sown with a chalk grassland seed mix.

The cultural heritage of the South Downs is also gradually being eroded through cultivation, neglect and the lack of funds for maintenance or restoration.

The past few decades have seen a succession of damaging changes:

- agricultural employment in the South Downs fell by 20% between 1985 and 1995[7];
- only 3% of the area remains as traditional chalk grassland[8];
- 90% of the precious lowland heaths have been lost (same ref as above);
- historic landscape features and archaeological remains are being lost due to clearance and ploughing;
- many ancient woodlands are in decline due to a lack of management;

4 Phase I Habitat Survey Data, derived from 1991 aerial photography.
5 The National Park Authority - a guide for members, ANPA, 2001.
6 FRCA figures, March 2000.
7 Hampshire Farming Study - a study of the viability of agriculture in Hampshire, Hampshire County Council, 1997.
8 From Rio to Sussex: action for Biodiversity, Sussex Biodiversity Partnership, 2000.

- the tranquillity of the area is being lost as a result of traffic noise and other intrusions.

The challenge will be to restore downland and some of the other habitats and wildlife, in partnership with land owners and managers, through positive land management and appropriate agricultural practices, linked to developing markets for the products of this landscape, such as lamb and charcoal.

To see the role that the Countryside Agency has suggested that a South Downs National Park Authority could play in land management, see page 25.

Leisure and recreation

The South Downs is already a heavily visited area, under significant recreational pressure from its mobile and leisure-conscious population. There is likely to be a rise in the demand for recreation as more people come to live and work in the area, whether or not the South Downs is designated as a national park.

- The area has a large resident population and a large and growing population around the proposed boundary - some 10 million people live within an hour's journey of the boundaries of the two AONBs, the Sussex Downs and East Hampshire.
- The population of the South East is expected to rise by around ten per cent over the next 20 years[9].

- A 10% rise is predicted in the demand for informal recreation within the area[10].
- A large proportion (77%) of visitors are local residents seeking a range of recreational activities[11].
- There is a substantial tourist industry based in the coastal towns that adjoin the boundary of the national park.

A challenge for a future national park authority will be to encourage quiet enjoyment of the national park while resisting recreational activities and developments that threaten its special qualities. Preserving tranquillity and landscape quality will be a significant task, but one for which a national park authority has the necessary skills and resources.

To see the role that the Countryside Agency has suggested that a South Downs National Park Authority could play in visitor management, see page 27.

A national park for the South Downs

The Countryside Agency believes that the South Downs meets national park criteria.

The proposed South Downs National Park is different in character to most of our existing national parks, which were created at a time when rugged open countryside with a sense of wilderness was central to designation. The proposed area, as well as being outstandingly beautiful is an exceptional

recreational resource of national importance. It is accessible, including by public transport, to a large urban population, and offers a sense of relative wilderness.

The Agency also believes that designation is the best way to provide for its needs. It would bring real and additional benefits to the conservation of the Downs that are not possible by any other management arrangements.

National park status would give the South Downs a permanent authority, committed to conserving and improving this highly valued landscape and able to address the pressures it faces in a comprehensive and integrated way. It would bring, for the first time, an independent body with the focused purpose of conserving the Downs; a permanent and unified body, forward looking, and able to plan strategically for the whole area and for the long term, with resources provided by government solely for this purpose. It would build on the valuable work that is carried out at present by the Sussex Downs Conservation Board and the East Hampshire Joint Advisory Committee.

The Countryside Agency is committed to creating the right national park for the South Downs. The rest of this document sets out our preferred options for the administrative arrangements for a South Downs National Park Authority and our first draft of where the national park boundary should be, and invites your views.

9 Dr Simon Woodward, South Downs National Park: Proceedings of the Countryside Agency Seminar, Countryside Agency, 2001.
10 Same as above.
11 SDCB, A Management Strategy for the Sussex Downs Area of Outstanding Natural Beauty, 1996.

Part A: Administrative arrangements
4   A South Downs National Park Authority

In this chapter we look at:
- how a South Downs National Park Authority would be set up;
- what a South Downs National Park Authority would do;
- how a South Downs National Park Authority would achieve its objectives.

We outline the administrative issues that need to be addressed:

Issue 1. Membership of a South Downs National Park Authority.

Issue 2. A role in forward planning and development control.

Issue 3. A role in land management.

Issue 4. A role in visitor management.

Issue 5. A national park management plan and delivery by the national park authority and others.

Issue 6. Working in partnership.

Issue 7. Involving local people.

For each, we:
- identify options for addressing the issue, explain the implications, and indicate the Countryside Agency's preferred approach at this stage; and/or
- indicate ways of working that a South Downs National Park Authority might wish to adopt.

We would like your views (see Chapter 1 for how to respond). Your response will be especially valued on these issues and options. However, please also comment on other issues you feel are relevant, bearing in mind the purposes of a national park authority. Remember, too, that this consultation considers administrative arrangements, not policy matters, which will be determined by the national park authority as and when it is established.

Part B considers the other main issue on which we would like your views, the draft national park boundary.

How a South Downs National Park Authority would be set up

Issue 1.
Membership of a South Downs National Park Authority

This section looks at who would be a member of a South Downs National Park Authority, how many members is should have, what experience and skills they should have and how some of them should be chosen. We look at the framework that is set out in the existing legislation, but there are other models. In Scotland, for example, legislation requires that the national park authority should have no more than 25 members, with at least 20% of its members directly elected by those living in the park and the remaining members appointed by the Scottish Minister. Of those, 50% are nominated by local authorities.

Membership framework

The current arrangements for national park authorities are set out in the National Parks and Access to the Countryside Act 1949 and the Environment Act 1995, see Appendix A) and are further explained in DoE Circular 12/96 (see Appendix C).

Under these arrangements, the members of a South Downs National Park Authority would be locally elected councillors and people with expertise in and knowledge of both the area and national park issues. It is likely to have around 46 members.

- One half plus one (24) would be elected councillors appointed to a national park authority by the local authorities (that is county and district councils and unitary authorities) whose areas lie within a national park. These appointees must be elected members of the local authorities, preferably members who represent a ward within the national park.
- The remaining members (22) are appointed by the Secretary of State.
- One half minus one (10) are parish representatives. Parish members must either be a member of a parish council, or chairman of the parish meeting, wholly or partly situated within the national park.
- One half plus one (12) are individuals appointed for their special expertise and experience, and to take account of the national park's purposes. In appointing individuals, the Secretary of State will look for people who have experience, preferably in a combination of fields, with direct relevance to the character of the particular national park. Wherever possible, preference will be given to candidates who combine these qualities and a local association with the national park.

The chairman of a national park authority is a member of it, elected by the members of that authority. The chairman may be either a local authority or Secretary of State appointee.

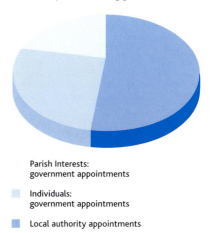

Parish Interests: government appointments

Individuals: government appointments

Local authority appointments

In setting up a South Downs National Park Authority, the Countryside Agency's preferred option for the size and structure of an authority is:

a) a full national park authority with around 46 members, as shown on the chart above.

This has the advantage that all local authorities would be involved and engaged in the process, and the full range of national and parish interests would be maintained. However, this would make a South Downs National Park Authority the largest national park authority in England and Wales. It would need to organise its business carefully to meet the Government's modernising agenda, which is for smaller decision-making bodies.
This could be, for example, through the use of executives/committees with delegated decision-making powers.

Another option could be:
b) a smaller South Downs
 National Park Authority.
This could be achieved without any change to the legislation if some local authorities agreed to be excluded, or if they agreed to reduce their representation. While this may be a more efficient way of conducting the authority's business, it could mean that some local authorities do not have a role. Because the overall proportions must remain the same, this would also mean fewer parish representatives and independent appointees.

Question 1A

Is the Agency's preferred option the right one for the South Downs?

Do you think the other option would be a better way forward, or are there any other options the Agency should consider?

Parish members

The arrangements for the selection of parish members vary between national parks. Usually, names are put forward to the Secretary of State from the parishes in the areas and he/she then appoints those suggested. **The Agency believes that there should be a locally agreed, open and democratic process for their selection, which will bring forward individuals from across the area and those with the appropriate knowledge and skills. Names would then be put forward to the Secretary of State for his appointment.**

One option would be to agree geographical areas within the national park as the basis for putting forward an agreed number of nominees (for example, by county) and holding local ballots among the parishes in each area to select preferred individuals (who must be parish councillors). The ballots could be organised by the County Associations of Town and Parish Councils.

Question 1B

How do you think parish members should be selected?

Creating a skilled administration

The Countryside Agency believes that members of a national park authority should have a clear understanding, knowledge and experience of the main issues that impact on the Downs.

National park authority members come from various backgrounds with differing levels of knowledge and expertise. The Countryside Agency believes that there should be a formal induction process for all national park authority members, followed by ongoing training. It also believes that there should be a code of conduct for all members, including a formal affirmation from members to achieving national park purposes in the South Downs. This would help to ensure that decisions made by all national park authority members meet national park purposes.

In order to make sure that members of the national park authority have the appropriate skills and knowledge, the Secretary of State could define the particular issues he believes members collectively should be able to address. This would then guide selection of his appointees to the national park authority, and also appointments by the local authorities and nominations by parishes. Advice could also be sought from expert or representational bodies, such as English Nature or the National Farmers' Union.

An alternative would be to have a larger proportion of individuals appointed for their particular expertise on the national park authority. Some of these may be appointed by other bodies.

**The Countryside Agency's preferred option is:
a) to keep to the present proportions of individuals and councillors, ie. 12:34 (see option 1a on page 17), but to give more guidance to local authorities and the Secretary of State as to the best balance of interests and expertise for authority membership. In addition, all members should receive training and sign a code of conduct that includes a commitment to national park purposes.**

The Agency thinks that the authority as a whole needs to have knowledge and experience in the following areas: farming, forestry, environmental

management, wildlife, cultural heritage, recreation, tourism, community participation and education.

An alternative could be:
b) new legislation to change the proportions, for example so that one half are local or locally appointed representatives, or are appointed by other bodies, and half are appointed by the Secretary of State. The higher proportion of individuals on the authority would be chosen for their relevant skills or expertise. However, this would mean a reduction in the number of elected and community representatives who bring their own expertise.

Question 1C

What are your views on the Countryside Agency's preferred option to ensure appropriate expertise is appointed to the authority?

Do the areas of knowledge and expertise set out cover the issues that affect the Downs?

Question 1D

Do you believe that mechanisms such as training and a code of conduct for members should be used to ensure that the national park authority is properly skilled?

Question 1E

What are your views on the other option?

Are there any other options the Agency should consider?

What a South Downs National Park Authority would do

Issue 2.
A role in forward planning and development control

This section looks at the role of a South Downs National Park Authority in forward planning and development control. Firstly we look at the nature of planning in the area and at the options for the preparation of development plans. We then consider the options for the administration of development control.

The planning process covers national and regional policy and guidance, including planning policy guidance (PPGs) and regional planning guidance (RPGs), development plans and development control (as determining planning applications in effect delivers development plan policy).

The role of development plans is to reconcile the demand for development with the protection of the environment, and thus provide the framework for rational and consistent decision making.

The Environment Act 1995 makes national park authorities the sole planning authority for the area. They have responsibility for preparing development plans for the national park area and for development control within their boundaries. This is because the ability to manage development - both to stop inappropriate

development, and also to use sustainable development as a positive force - is central to achieving national park purposes. Legislation allows the Secretary of State to transfer plan-making powers (but not development control) to other local authorities. However, this has not happened in any other national parks.

The particular characteristics of the South Downs mean that the planning arrangements that operate in other national parks may not necessarily be the best for the South Downs. The Agency has looked at best practice in other English and Welsh national parks and considered the scope offered by the existing legislation and any need for change. There is a wide range of options for the administration of planning in a South Downs National Park. These need to take into consideration the needs of a national park, as well as the implications for the existing planning authorities, for cost and efficiency, and for users of the planning system.

Being responsible for development plans and development control gives a national park authority the ability to protect the national park from inappropriate development and so carry out its statutory duties. It ensures close links between the planning system and the national park management plan and so enables forward planning and development control to be implemented to achieve national park purposes in a positive way.

The nature of planning in the South Downs differs from other existing national parks, because of the area's characteristics, which include:

- the area's shape (approximately 116 miles long and only 6.5 kms across in places);
- its juxtaposition with major settlements, in particular the south coast conurbations sandwiched between the sea and the Downs;
- the north-south transport corridors;
- the number of local authority areas it covers;
- its location in the populous South East, with severe development pressures and consequential development control caseload.

The combination of these factors presents particular challenges for establishing a new, park-wide planning authority. An effective planning system for the South Downs must ensure:

- certainty in achieving national park purposes;
- cross-boundary cooperation between the park and neighbouring authorities;
- the ability to deal efficiently with the volume of the development control work, much of which will be small-scale and urban in nature;
- accessibility for users of the planning system across the whole area;
- efficient use of resources and minimal duplication of existing arrangements.

Regional planning guidance

Regional planning guidance sets the framework for development plans. It should recognise the national importance of the South Downs and set policies in line with national guidance such that the national park is given the highest level of protection. The Countryside Agency believes that the national park authority should have a strong influence on regional planning guidance, which will set the regional context for planning in and around the national park.

Structure plan

The structure plan, which sets out the strategic policies and provides a framework for local plans, must conform to national and regional planning guidance.

Government guidance (in DoE Circular 12/96, see Appendix C) encourages joint structure planning for national parks. Most of the existing national parks have chosen to enter into voluntary joint arrangements with neighbouring county councils or unitary authorities in preparing structure plans. Joint structure planning in the South Downs would go some way towards addressing the key question of strategic cross-boundary issues, and would set a coherent policy framework for all local plans in and around the national park.

The Countryside Agency believes that a shared and joint approach to structure plans is therefore best for the South Downs. The size of the area,

however, means that one single joint plan for the national park and the three surrounding counties would be too large and complex to be workable. It would involve seven structure plan authorities - covering an area from Kent to Dorset - working together and to the same timescale.

The Countryside Agency's preferred option is therefore:

a. **joint structure plans prepared together by a South Downs National Park Authority and the existing structure plan authorities[12] for the three county areas, and adopted by both the national park authority and the existing structure plan authorities.**

Given the geography and administrative complexity of the South Downs, there should be a series of three (one per county) joint structure plans. (This could be implemented by the Secretary of State using secondary legislation under the Environment Act 1995.) The national park authority would retain legal responsibility for the structure plan for the national park, and common strategic policies for the national park area would appear in each plan. This would allow strategic consideration of cross-boundary issues, for example housing and transport routes. The national park authority could also influence policies in the surrounding area which might affect the park.

---

12 Hampshire County Council, Southampton City Council, Portsmouth City Council, West Sussex County Council, East Sussex County Council and Brighton & Hove City Council.

Other options are:

b. that the national park authority prepares a structure plan on its own. This could be achieved to some extent by close working between neighbouring authorities, for example through joint advisory committees. However, this reduces the opportunity for a joint approach to strategic cross-boundary issues and for a common framework for all local plans in the area;

c. a joint structure plan with one of the existing structure plan authorities (for example West Sussex, as the county with the largest area in the national park). This would have the benefits set out above for one of the three counties, but would leave difficulties in the remaining two counties in adopting a common approach;

d. a legal transfer of responsibility from the national park authority to the existing authorities for structure plan preparation for the park area, together with the rest of the county - but this would leave the national park authority with no role in the strategic plan, reducing the certainty of achieving national park purposes. It could also mean no national park context for the local plan(s) and so would weaken links with the national park management plan.

Question 2A

What are your views on the Countryside Agency's preferred option for a South Downs National Park Authority to prepare joint structure plans?

Question 2B

Do you think one of the other options would be a better way forward, or
are there other options the Countryside Agency should consider?

Minerals and waste local plan

Minerals and waste plans set out local authorities' detailed land use policies for the management and disposal of waste, within the broad strategic framework of the structure plan. The plans address the need for sites and facilities in particular areas, suitable locations, and the planning criteria likely to apply. They also carry forward policies that provide for the supply of minerals and for ensuring the required degree of environmental protection associated with development. They can also set out the development control criteria that will be applied in considering applications for mineral working and requirements for the restoration and aftercare of such sites.

Issues relating to minerals and waste local plans are similar to those for structure plans in that they require a shared strategic approach across boundaries, including both towns and the countryside.

The Agency's preferred option is therefore:

a. **that minerals and waste local plans should be prepared jointly with the existing minerals and waste planning authorities[13] for the three county areas, as this is a strategic matter that should be dealt with using a joint approach across boundaries (as for structure plans).**

The other options are:

b. that the national park authority should prepare minerals and waste local plans on its own. They could either prepare this separately or as part of a park-wide local plan;

c. a full transfer of responsibility from the national park authority to the existing authorities for minerals and waste local plan preparation for the park area, together with the rest of the county.

Question 2C

What are your views on the Agency's preferred option for three joint minerals and waste local plans?

Question 2D

Do you think one of the other options would be a better way forward, or are there other options the Agency should consider?

13 The same as the structure plan authorities.

Local plan

Legislation makes the national park authority responsible for local plans, so in existing English national parks, the national park authority prepares its own park-wide local plan. The purpose of a local plan is to establish a comprehensive set of land-use policies for the area. It guides development within the plan area and provides a basis for determining planning applications. It must conform with structure plans, and also to central government regulation and Department for Transport, Local Government and the Regions' circulars and PPGs, as well as regional planning guidance.

The issues relating to structure planning in the South Downs also apply to local plans. It will be important to make links between the national park area and those areas beyond. It will also be important to have coherent policies across the park, which give details about development sites and standards to ensure common decisions to the highest possible standards. These policies must also be closely linked to the national park management plan, so that planning does not simply allow or restrict development, but achieves sustainable development that contributes positively to national park purposes.

Although there are some strong arguments for joint local plans, as there are for structure plans, the administrative complexity and the risk of duplication of effort for a national park authority working with 12 local plan authorities would make this extremely difficult. It would be especially challenging to achieve a consistent park-wide approach.

The Countryside Agency's preferred option is:

a. a park-wide local plan prepared by the national park authority.

The national park authority should address cross-boundary matters by working in conjunction with the neighbouring authorities, for example through a joint advisory committee. (The Secretary of State could direct the national park authority to pay due regard to the advice of other authorities in preparing its plan.) The national park local plan would be prepared in conformity with a joint structure plan(s) for the area.

The benefits of this arrangement would be to ensure a consistent approach to land use planning across the park, in line with national park purposes. It will also allow a robust policy framework for national park purposes when taking development control decisions, together with close links between land use planning and the national park management plan.

Other options are:

b. the national park authority prepares the plan by itself, with no special arrangements to consult its neighbours. This, however, does not address the cross-boundary issues pertinent to the South Downs, given the shape, geography and pressures upon it;

c. a series of up to 12 joint local plans - this would be extremely inefficient for the national park authority, which would prepare 12 instead of one plan at different times, and would risk inconsistencies of policy across the park area;

d. legal transfer of responsibility from the national park authority to the existing local planning authorities. This would, however, mean that a South Downs National Park Authority had no role in local planning for the area and would therefore undermine its ability to ensure that national park purposes (particularly conservation) are achieved. It would also risk inconsistencies of policy across the park area.

Question 2E

What are your views on the Countryside Agency's preferred option for a South Downs National Park Authority to prepare a park-wide local plan, working in conjunction with constituent and neighbouring local authorities?

Question 2F

Do you think one of the other options would be a better way forward, or are there other options the Countryside Agency should consider?

Unitary development plans

An alternative to the options set out above for structure, minerals and waste and local plans would be for the national park authority to prepare a single unitary development plan. This would encompass all three development plans (but not the local transport plan). This could be more efficient and the Secretary of State could enable the national park authority to do this under the Environment Act 1995. However, this would not bring the advantages of joint planning that are set out in the sections above.

Question 2G

Do you think that a unitary development plan would be the best model for the South Downs?

Development control

Development control (ie making decisions on individual planning applications) is the responsibility of all English national park authorities, under the provisions of the Environment Act 1995. Elements of this responsibility can be shared with or delegated to other local authorities (but cannot be legally transferred to other local authorities under existing legislation).

A South Downs National Park Authority is likely to have far more development control casework than other national parks. Current statistics suggest in excess of 2,500 planning applications per year in the proposed national park area, compared with an average of around 600 in other national parks. Many of these are likely to be individual householder applications and to arise in towns and settlements and so would have limited impact on national park purposes. The Agency has therefore considered carefully how this workload might be handled most effectively. There may be benefits from local authorities carrying out some development control work in a national park, working to policies and standards defined by a South Downs National Park Authority. This could apply to both minerals and waste applications, currently handled by county councils, and other applications currently handled by the district councils.

This would make good use of the technical expertise and local knowledge of the existing planning departments (especially in the specialist area of minerals and waste planning). It would be cost effective because it would make use of existing planning staff, and so the national park authority may not need to set up a large planning team of its own. It would also help make strong links between planning inside the park and outside its boundaries and contribute to sustainable development for the area as a whole. In addition, it would help ensure an effective and efficient system for users, who could make and comment on applications through local council offices, for example.

A legal transfer (under new primary legislation) of development control powers from a South Downs National Park Authority to the existing development control authorities would mean that a national park authority would have no say in this important part of the planning process. There would therefore be a risk that applications could be approved that conflict with national park purposes. It would also reduce the opportunity to make links between development control and the national park management plan; so reducing the opportunities for sustainable development and the environmental, social and economic benefits this can bring to achieve national park purposes.

However, some elements of the system could be delegated (ie carried out by another body by local agreement) so that the existing councils would carry out development control on behalf of a South Downs National Park Authority. They would work to policies prepared by the park authority in the national park (joint) structure and local plans.

The national park authority would remain responsible overall for development control. Delegation would be agreed to so that a South Downs National Park Authority would retain the powers to determine any applications that could conflict with national park purposes or contradict the local or structure plans, or where cases are contentious or significant. Delegation would need close working between all local authorities and the national park authority to share expertise and to ensure that decisions were consistent throughout the national park and in line with the national park management plan (for example, on design standards).

Decisions on delegation would be a matter for negotiation between a South Downs National Park Authority and the local authorities, and would be subject to agreement from each of them. If agreement could not be reached, the responsibility would remain with the national park authority. However, the Secretary of State can give guidance to the national park authority and local authorities, and the Countryside Agency can advise, on how the national park authority should carry out its functions.

An alternative could be that if the national park authority itself retained all responsibility for development control, it could set up local area sub committees which could seek advice from neighbouring local authorities. This mechanism could be used for the cases that the national park authority determines, even if some responsibility were delegated to existing councils.

The Countryside Agency's preferred option is:
a. that the national park authority should retain responsibility for development control, but should delegate casework to local authorities, which would make decisions in line with structure and local plan policies set or agreed by a South Downs National Park Authority.

The national park authority should retain the right to make decisions on major or contentious cases, or those out of line with policy. The degree of delegation would be negotiated and agreed between a South Downs National Park Authority and the relevant authorities. It might range from case paperwork being handled by officers, through minor decisions being taken by local authority officers, to decisions being taken by local authority planning committees.

Other options could be:
b. no delegation, with a South Downs National Park Authority dealing with all aspects of development control, perhaps using local area committees. This may lead to inefficiencies, with both the national park authority and the local authorities running development control departments, and time being spent by the national park authority on dealing with cases that have little or no impact on national park purposes;

c. a legislative amendment to allow transfer of development control from a South Downs National Park Authority to the local authorities. This would, however, leave the national park authority with no statutory role in the process and result in its having no influence, even over major developments that would have a significant impact on national park purposes. This would mean that the authority would be unable to achieve its statutory purpose.

Question 2H

What are your views on the Countryside Agency's preferred option to delegate some development control responsibilities to existing local authorities?

What are your views on the degree of delegation that would be appropriate?

Question 2I

Do you think one of the other options would be better, or are there any other options the Countryside Agency should consider?

Issue 3.
A role in land management

In this section we look at the role the Countryside Agency believes that a South Downs National Park Authority should play in land management. First we look at land management in general, then we consider farming and forestry, nature conservation and cultural heritage.

The land in the South Downs is primarily privately owned and managed, and this would not be changed by national park designation. However, a national park authority could encourage landowners to work towards national park objectives in managing their land.

How land is looked after and managed is crucial to the maintenance of the unique landscape and natural beauty of the South Downs, and therefore to the conservation of both the natural environment and cultural heritage.

The dramatic landscape of the South Downs is a complex and diverse lowland landscape of downland, river valleys, woodland, heathland and traditional and more intensive agriculture. This landscape has been shaped by the interplay between the physical influences of geology, landform, climate and human activities. The continuation of this interplay and the role of landowners and managers is vital to the future of the Downs.

This is a living landscape, one where people live and work. The South Downs will remain predominantly a landscape of mixed farming and forestry. Its management must respect and reflect this, through appropriate agricultural and forestry practices.

A South Downs National Park Authority would have a vital role to play in not only conserving the landscape, heritage and rural economy, but also in seeking to restore some of the important landscapes and habitats that have been lost over the years. A South Downs National Park Authority would act as a focal point for existing agencies and schemes, bringing them together to act in a more integrated and effective way. In particular, the

Countryside Agency believes that it should:
- **develop a vision for the long-term needs of the area and a coherent set of land management policies to address those needs;**
- **take a strategic view of the land management resources of the national park, monitoring change and seeking solutions;**
- **take a lead role in co-ordinating policy and setting priorities, by working closely with other statutory and voluntary bodies, particularly the Department for Environment, Food and Rural Affairs.**

Farming and forestry

The South Downs is largely an agricultural landscape. Over 57% of the area is in arable cultivation. This is a higher percentage and overall area than any other national park. This is not a recent development, but has been the case for centuries: woodland clearance was almost total by the beginning of the Roman occupation (except for the Weald, which was still wooded), with arable cultivation then at levels not exceeded until after the Second World War.

The traditional mixed pattern of downland farming used a range of landscape types, from chalk uplands to river valleys and floodplain pastures. In the past 50 years, however, intensification of agricultural practices has resulted in the loss of this pattern.

About 24% of the area (approximately 27,000 hectares)[14] within the two AONBs is covered by woodland. Most of this is privately owned. About a third has been planted and two-thirds is of ancient origin (ie. has remained wooded for more than 400 years), much of which is in decline due to lack of management. Trees that are poorly managed fail to achieve their potential growth rate, and neglect results not only in loss of income but also in a reduced wildlife value.

In the 1980s, efforts were made to halt the decline in traditional habitats, by encouraging farmers to manage their land in more

---

14 Phase I Habitat Survey data, derived from 1991 aerial photography.

environmentally sensitive ways. Much of the chalk landscape of the Downs was designated as an Environmentally Sensitive Area (ESA)[15] in 1987, and elsewhere the Countryside Stewardship Scheme offered voluntary agreements to farmers to positively conserve areas of their land.

Such efforts stemmed, but did not reverse the decline. Species-rich chalk grassland, for example, now covers less than 3% of the chalk landscape of the Downs. The potential to expand and enhance the chalk grassland resource is considerable. The Sussex Biodiversity Plan for chalk grassland aims to recreate 1,000 hectares of chalk grassland by 2010. However, reversion back to a species-rich chalk grassland is a slow process and results are variable. The restoration of the landscape requires an integrated approach, needing not only to meet environmental objectives, (for example the creation of 1,000 hectares of chalk grassland), but also to meet social and economic objectives as well: where will the sheep come from to graze these new areas? are there markets for the lamb? which skills are needed to look after livestock compared with arable farming?

The Agency believes that a South Downs National Park Authority should:
- **work with the Department for Environment, Food and Rural Affairs and other bodies to ensure that the existing agri-environment schemes in the area, particularly the English Rural Development Programme (which includes the ESA and Countryside Stewardship Schemes) are well targeted;**
- **run its own intergrated rural development scheme to encourage landscape restoration and sustainable farming. This would require a significant and specific budget, joint working and some delegation from the Department for Environment, Food and Rural Affairs to the national park authority;**
- **offer a park-wide integrated countryside management service, providing a first-stop-shop advisory service for farmers and landowners;**
- **provide a clear mechanism for discussing and acting upon farming and forestry issues, for example through panels, a forum or working groups;**
- **work with woodland owners to develop integrated management and conservation plans for woodlands, particularly in respect of smaller holdings;**
- **promote and support initiatives to market local produce;**
- **influence agri-environmental policy at a national and international level.**

Nature conservation

Centuries of traditional land management practices have resulted in a rich diversity of natural habitats, including chalk grassland, ancient woodland, flood meadows, lowland heath, scarce chalk heathland and coastal habitats[16]. However, although 12,000 hectares of land are currently protected by national and international nature conservation designations, these traditional habitats have continued to suffer serious decline. The loss of important habitats through destruction or lack of management, particularly since 1945, has had severe consequences for flora and fauna in the Downs. The biologically important habitats that remain are small, fragmented, frequently under-managed and at risk.

Cultivated land (arable and pasture) is also important for wildlife. However, agricultural intensification has greatly depleted species diversity, such as farmland birds and arable flowers. The remaining wildlife interest is mainly confined to field margins and hedgerows. Some species, such as the stone curlew, have become extinct from the South Downs during this century and others, such as the Adonis Blue and Duke of Burgundy butterflies, are threatened.

15 The ESA was designated by the Ministry of Agriculture, Fisheries and Food, now the Department for Environment, Food and Rural Affairs.
16 There are significant areas of floodplain grassland, of international importance to nature conservation, within the floodplains of the Rivers Itchen and Arun.

The Countryside Agency believes that a South Downs National Park Authority should:

- seek to implement relevant local and national Biodiversity Action Plans through land management policies and activities;
- work closely with other organisations, such as English Nature, the local authorities and the county wildlife trusts, through memorandums of understanding and accords;
- provide a clear mechanism for discussing and acting upon nature conservation issues, for example through panels, a forum or working groups.

Cultural heritage

Past human exploitation of the land influences landscapes that today form the character of the area. The South Downs has long provided a home to settlers, and this history of interaction with the landscape has left a rich archaeological and cultural legacy. There are remains from every period - from the earliest Bronze Age settlements and Iron Age hill forts, to Roman roads, deserted medieval villages, great Tudor houses and fine historic parkland landscapes.

The proposed national park area has many important literary and artistic associations too. Its stunning landscapes have inspired generations of writers and artists, including Rudyard Kipling, Hilaire Belloc, Virginia Woolf, Gilbert White, Edward Thomas, Edward Elgar, Turner and Constable. Cultural influences are as strong today, and the area's unique sense of place continues to find expression through the work of writers, artists and musicians.

There has always been change within the Downs, but it is the great increase in the scale and pace of development, especially in agriculture and forestry, during the past 50 years that has threatened much of the remaining physical evidence of the past. Much of the area's rich cultural heritage is vulnerable, and is being gradually eroded (particularly the smaller features not subject to planning controls) through neglect and a lack of funds for restoration and maintenance.

The Agency believes that a South Downs National Park Authority should:

- play an active role in the conservation of cultural heritage;
- work with other organisations, such as English Heritage and the local authorities, through accords and agreements;
- provide a clear mechanism for discussing and acting upon cultural heritage issues, for example through panels, a heritage forum or working groups.

Question 3A

What are your views on the proposed role for a South Downs National Park Authority in relation to farming and forestry, nature conservation and cultural heritage?

Do you agree that restoration of downland should be a particular priority?

Question 3B

Are there any other land management matters that a South Downs National Park Authority would need expertise to address?

Issue 4.
A role in visitor management

This section looks at the role the Agency believes a South Downs National Park Authority could play in providing a countryside management service, in managing rights of way and access and in related activities such as transport, tourism, education and interpretation.

The South Downs are already well known and well loved (one survey estimates that the Sussex Downs receives in excess of 32 million visits a year, more than many national parks[17]). Designation would not necessarily generate an increase in existing visitors (although numbers will grow as a result of the growing population in the South East and the number of houses likely to be built). However, national park status would bring new resources, which should mean that the South Downs can be more

17 Osborne, Quantifying the amenity value of the Sussex Downland, University of Sussex, 1995.

effectively managed for people to enjoy them without causing more damage, for example by providing better public transport.

The national park authority would promote the understanding and enjoyment of the area. Its role is not to promote recreational activity as such, nor to increase the number of visitors, but to manage it in ways that minimise negative impacts on both the unique environment and on the local community. It would also work to harness positive outcomes, such as income for local business.

There are already many bodies and individuals involved in visitor management in the South Downs with whom a South Downs National Park Authority would work.

A countryside management service and site management

At present there is no single provider of countryside management services covering the whole area. Countryside management involves rangers and volunteers in managing sites and rights of way, as well as providing advice and information to landowners, visitors and the local community. In Sussex, a countryside management service is currently provided by the Sussex Downs Conservation Board and within the East Hampshire AONB, by Hampshire County Council, working with AONB staff. A number of key sites within the draft national park boundary are owned and managed by local authorities[18].

The Countryside Agency believes that:
- **the South Downs National Park Authority should provide a fully integrated, area-based countryside management service;**
- **a South Downs National Park Authority should develop a framework that ensures common high standards for the management of publicly owned land, including visitor sites, so that they meet national park purposes;**
- **where appropriate, a South Downs National Park Authority should manage (on behalf of the local authority or private owners) sites such as country parks that play an important strategic role in extending and improving access to the national park and helping people to understand and enjoy the Downs;**
- **a South Downs National Park Authority should consider owning such land itself, where this is the most effective way to create new public access.**

Question 4A

Do you think that a South Downs National Park Authority should run its own integrated and area-based countryside management service?

Are there other options that the Agency should consider?

Question 4B

What do you think a South Downs National Park Authority's role should be in relation to site management and ownership?

Access and rights of way

A South Downs National Park Authority would have legal responsibilities for managing access to open country and common land under the provisions of the Countryside and Rights of Way Act 2000. The national park authority would have powers to:
- operate a Local Access Forum, under the terms of the Act;
- make byelaws to preserve order, to prevent damage on access land and to avoid undue interference with the enjoyment of the land by others;
- appoint rangers in respect of access land;
- erect notices indicating the boundaries of access land and excepted land, notifying the public of any restrictions or exclusions in force and providing general information, and to contribute towards the cost of such signs provided by anyone else;
- seek agreements with landowners for the creation or safeguarding of means of access to access land;
- require landowners to provide means of access to access land.

18 For example, Queen Elizabeth Country Park is partly owned and managed by Hampshire County Council, East Sussex County Council owns Seven Sisters Country Park, which is managed by the Sussex Downs Conservation Board. Eastbourne Borough Council currently owns and manages land around Beachy Head.

As a relevant authority, the national park authority will have the duty to manage the imposition of all exclusions and restrictions imposed on access land within the national park boundary. (Outside the national park boundary the duty falls to the Countryside Agency.)

However, as national park authorities are not highway authorities they are not responsible for managing rights of way, although in many national parks the highway authority has delegated some responsibility to them. Government guidance in DoE Circular 12/96 (see Appendix C) encourages them to do this.

Compared with other national parks, within the South Downs there is likely to be only a relatively small amount of land that is classified as open country under the Countryside and Rights of Way Act. With the exception of the Broads Authority, where there is an extensive system of public navigation, the English national parks have between 40,000 and 103,000 hectares of land that is classified as open country. Within the Downs the figure has been estimated at about 7,700 hectares. Around 11,000 hectares of land is currently accessible to the public through existing access agreements. This includes, for example, land owned by the Forestry Commission, country parks and land in Countryside Stewardship Schemes with access agreements.

There is, however, an extensive rights of way network, with over 4,000 km of footpaths and bridleways throughout the Downs, including the South Downs Way National Trail. These rights of way provide people's main form of access in the national park, and will continue to do so. The Countryside Agency therefore believes that a South Downs National Park Authority should have a lead role to play in planning for and managing access and rights of way.

The Countryside Agency's preferred option is:

a. that a South Downs National Park Authority should have overall responsibility for rights of way within the national park. Powers and resources should be delegated to them by all highway authorities to enable the national park authority to be responsible for:

- **keeping the definitive maps under review;**
- **making orders altering public rights of way;**
- **maintaining the rights of way network;**
- **producing and implementing the Rights of Way Improvement Plan.**

This would not necessarily mean that the national park authority carried out all these operations itself - it could, for example, use existing specialist expertise in local authorities, but it would mean the national park authority would be able to take decisions on the quality and quantities of rights of way. This would result in common, high standards throughout the national park.

Other options are:

b. that the national park authority is responsible for managing the rights of way but does not act as the highway authority. This could mean that rights of way are well managed but that the national park authority would not be responsible for producing the Rights of Way Improvement Plan and would not be able to make any improvements to the network. This would also mean that people would have to contact more than one organisation about rights of way and access issues within the national park;

c. that the highway authorities should continue to be responsible for rights of way management. This would mean, however, that the national park authority would have no influence over the most significant recreational resource in the national park.

Question 4C

What are your views on the Countryside Agency's preferred option for highway authorities to delegate rights of way powers to a South Downs National Park Authority?

What activities do you think the national park authority should be responsible for?

Question 4D

What are your views on the other options, or are there any other options that the Agency should consider?

Transport

The proposed national park is close to large centres of population, and is served by a good road network. A large proportion of those who visit it, even if only on a short journey, will do so by car. Many of these journeys are to a limited range of sites or areas within the Downs, often using small country lanes. Some popular tourist sites also suffer traffic problems because of high visitor numbers. The proximity of urban areas on the periphery of the South Downs creates problems, with 'rat runs' for commuter traffic and lorries impacting on local communities and tranquillity.

Local authorities would remain as the highway authorities with responsibility for transport and traffic management, including the production of Local Transport Plans. These five-year strategies set policies for traffic management and planning and managing the highway network and rural transport.

Local Transport Plans are produced in consultation with local communities and partners, including a national park authority. A South Downs National Park Authority might seek to ensure, for example, that there is sufficient access to quiet roads for activities such as cycling and horseriding.

The impact of cars on tranquillity and enjoyment of the Downs is high, and there is clearly a need to manage traffic. In comparison with other national parks, the South Downs benefit from a good public transport network. Therefore, greater use of public transport would reduce vehicle pressure. A South Downs National Park Authority could develop initiatives to encourage its use.

The Countryside Agency believes that a South Downs National Park Authority should play an active role in:

- **the preparation of Local Transport Plans, working jointly with the highway authorities on policies in the national park, or for surrounding areas that impact upon it;**
- **delivering, using its own powers, parts of the transport strategy developed by the highway authorities, particularly where related to achieving sustainable recreation, for example by supporting local bus services;**
- **implementing transport policy (traffic management, for example) on behalf of the highway authority through delegation of powers. The highway authority should provide resources (including funding) to the national park authority through the Local Transport Plan, to allow it to act on its behalf.**

Question 4E

Do you agree that the national park authority should have an active role in transport and traffic management?
Are there other issues that the Countryside Agency should consider?

Tourism

A South Downs National Park Authority would not be the tourism authority, and would not promote the South Downs to tourists. This role would continue to be undertaken by the Tourist Boards and local authorities. The national park authority would, however, need to work closely with those who do cater for visitors and the tourist authorities to ensure that tourism does not conflict with national park purposes.

The Countryside Agency believes that:

- **a tourism strategy should be prepared jointly between the tourist authorities and the national park authority.**

Question 4F

Do you agree with the role outlined for the national park authority on tourism?
Are there other issues that the Countryside Agency should consider?

Education and interpretation

National park purposes mean that provision of information would be an important task for a South Downs National Park Authority. It would, for example, provide the lead in developing the full range of interpretive provision within the national park, and would promote best practice. Such activities would take place in partnership with a range of other agencies, with the authority ensuring the delivery of high-quality, appropriate interpretation across the proposed national park area.

A South Downs National Park Authority would also work in partnership with the education sector, developing resources and materials for schools and other groups.

The Countryside Agency believes that the national park authority should:

• **develop and coordinate interpretation of the national park, for example through a South Downs Interpretation Forum;**

• **agree, with local authorities and others, a shared interpretative strategy for publicly owned land and other important sites, to promote common messages;**

• **promote community visitor information sites, such as shops and post offices, and improve existing facilities;**

• **develop local outreach programmes to encourage use of the Downs by schools and groups of young people, and develop a research agenda in partnership with the further and higher education sector and local communities.**

Question 4G

Do you agree with the role outlined for the national park authority on education and interpretation?
Are there other issues that the Countryside Agency should consider?

How a South Downs National Park Authority would achieve its objectives

Issue 5.
A national park management plan and delivery by the national park authority and others

The Environment Act 1995 requires each national park authority to prepare and publish a national park management plan. This acts as an umbrella document for the authority's work, being a statement of the authority's policies for managing the park and for carrying out its functions. In practice, the plan provides the strategic framework and policies for the park as a whole.

The Act also places a duty on all relevant authorities[19] to take account of national park purposes when making decisions or carrying out their responsibilities within a national park.

National park management plans do not form part of the development plan system. They concentrate on the primary purposes of conserving and enhancing the natural beauty,

wildlife and cultural heritage of the area and promoting opportunities for the understanding and enjoyment of the special qualities of the area by the public. Their function is to coordinate not only the work of the authority itself but also of other agencies and partners, including both the private and voluntary sectors.

The plans are subject to wide consultation during their preparation. Their implementation involves individuals as well as local and national organisations. The plans are reviewed every five years, which allows monitoring of achievements and ensures that there is ongoing consultation and that changing circumstances can be reflected.

The Countryside Agency believes that:

• **the national park management plan's role in influencing the policy and programmes of the national park authority and other bodies, as they relate to the objectives of national park designation, should be confirmed. This could be achieved by the Secretary of State issuing guidance on the matter, for example through a circular;**

• **a South Downs National Park Authority should monitor its own role in implementing the national park management plan against annual targets, and should**

19 Relevant authorities are defined as any Minister of the Crown, any public body, any statutory undertaker or any person holding public office.

audit the performance of other statutory agencies and local authorities in achieving agreed policies within the plan;
• public bodies should incorporate within their own business plans a clear statement of how they will fulfill their duty to take account of national park purposes when making decisions or carrying out their responsibilities within the national park.

Question 5A

What are your views on the role of the national park management plan, and the role of a South Downs National Park Authority in co-ordinating and monitoring action by others?

Are there any other issues that the Countryside Agency should consider?

Issue 6.
Working in partnership

A national park sits within a wider social, economic and environmental context: locally, regionally and nationally. In order to achieve national park purposes in the South Downs, an authority would need to work closely with local authorities, and with statutory and voluntary bodies within and around the national park boundary, and across the region and beyond.

The Countryside Agency believes that effective partnership working is best achieved through strategic alliances and innovative approaches. Such approaches include secondments and staff exchanges, as well as local memoranda of agreement or accords. It might, for example, involve the national park authority:
• entering into accords with public bodies which have a role to play in achieving national park purposes. These might include: the Department for Environment, Food and Rural Affairs, in its role of implementing the English Rural Development Programme; South East Economic Development Agency, in its role of promoting sustainable development; as well as English Nature, English Heritage and others;
• seeking opportunities for national park members to serve on other bodies or committees, for example by securing a place on the South East Regional Assembly.

It is also important that a South Downs National Park Authority becomes part of the family of protected areas in the UK and internationally, sharing good practice and gaining from the experience of other national parks.

The Countryside Agency believes that the national park authority should:
• develop working relationships with other national park authorities and with bodies that manage other protected areas, such as the Association of National Park Authorities and Europarc (the Federation of Nature and National Parks for Europe);
• form links with voluntary bodies, such as the wildlife trusts, the National Trust and the Council for National Parks.

Question 6A

What are your views on involving other parties through joint working in order to support the work of a South Downs National Park Authority?
Are there any other options that the Countryside Agency should consider?

Issue 7.
Involving local people

A South Downs National Park Authority would need to involve and consult its local communities and visitors, to take account of the wide range of interests that can be affected by their work. These include the interests of those who live and work in the national park, and those who derive their living from its resources. It should also involve them actively in decision making and in the formulation of policies and programmes. In doing so, the authority should seek to be as inclusive as possible, by involving people within the boundary and those living outside the park. It should also seek to ensure that all parts of the community, including young people, are given a voice in the way the national park is run. It also needs to develop particular arrangements to inform, consult and involve parish and town councils in the running of a national park authority (see DoE Circular, Appendix C).

The Government's Rural White Paper[20] proposes that communities could play a much bigger part in running their own affairs. Part 1 of the Local Government Act 2000 places a duty on local authorities to produce 'community strategies'[21] for the promotion or improvement of the economic, social and environmental well-being of their areas. A national park authority would have a key role in working with local communities, and therefore would be well placed to contribute to the development and implementation of such strategies, and to reflect them in its own plans and actions. Equally, the national park management plan should contribute ideas to community strategies.

The Countryside Agency believes that the national park authority should put in place arrangements which will actively involve local people. For example the national park authority could:

- **hold a public forum or general meeting at least annually - it should also hold public events for local people to participate in when deciding key issues, for example in preparing the national park management plan;**
- **normally conduct its business in open meetings at which the public is given an opportunity to participate;**
- **communicate clearly and on a regular basis with local people and visitors, for example through consultations, newsletters and a website;**
- **create partnerships with education authorities, and with user and community groups, to develop opportunities to increase public understanding of the special characteristics and conservation needs of the national park;**
- **establish a children's advisory committee or board for the national park authority, to ensure that young people understand and make best use of the Downs on their doorstep and have a say in the area's future;**
- **work closely with the local authorities in the preparation and implementation of local authority community strategies.**

In light of the number of communities within the proposed national park, a South Downs National Park Authority will need to develop a close relationship with town and parish councils, and community groups, as well as with the county and district associations of local councils, rural community councils (or their equivalent), and local amenity societies.

Question 7A

What are your views on how a South Downs National Park Authority might involve local people?
Are there any other issues that the Countryside Agency should consider?

20 DETR and MAFF, Our Countryside: the future. A fair deal for rural England, 2000.
21 Also known as community plans.

5 Summary of administrative issues and questions

Issue 1. Membership of a South Downs National Park Authority

1A. Is the Agency's preferred option the right one for the South Downs?
Do you think the other option would be a better way forward, or are there any other options the Countryside Agency should consider?

1B. How do you think parish members should be selected?

1C. What are your views on the Countryside Agency's preferred option to ensure appropriate expertise is appointed to the authority? Do the areas of knowledge and expertise set out cover the issues that affect the Downs?

1D. Do you believe that mechanisms such as training and a code of conduct for members should be used to ensure that the national park authority is properly skilled?

1E. What are your views on the other option? Are there any other options the Agency should consider?

Issue 2. A role in forward planning and development control

2A. What are your views on the Countryside Agency's preferred option for a South Downs National Park Authority to prepare joint structure plans?

2B. Do you think one of the other options would be a better way forward, or are there other options the Countryside Agency should consider?

2C. What are your views on the Agency's preferred option for joint minerals and waste local plans?

2D. Do you think one of the other options would be a better way forward, or are there other options the Agency should consider?

2E. What are your views on the Countryside Agency's preferred option for a South Downs National Park Authority to prepare a park-wide local plan, working in conjunction with constituent and neighbouring local authorities?

2F. Do you think one of the other options would be a better way forward, or are there other options the Countryside Agency should consider?

2G. Do you think that a unitary development plan would be the best model for the South Downs?

2H. What are your views on the Countryside Agency's preferred option to delegate some development control responsibilities to existing local authorities?

What are your views on the degree of delegation that would be appropriate?

2I. Do you think one of the other options would be better, or are there any other options the Countryside Agency should consider?

Issue 3. A role in land management

3A. What are your views on the proposed role for a South Downs National Park Authority in relation to farming and forestry, nature conservation and cultural heritage? Do you agree that restoration of downland should be a particular priority?

3B. Are there any other land management matters that a South Downs National Park Authority would need expertise to address?

Issue 4. A role in visitor management

4A. Do you think that a South Downs National Park Authority should run its own integrated and area-based countryside management service?

Are there other options that the Agency should consider?

4B. What do you think a South Downs National Park Authority's role should be in relation to site management and ownership?

4C. What are your views on the Countryside Agency's preferred option for highway authorities to delegate rights of way powers to a South Downs National Park Authority? What activities do you think the national park authority should be responsible for?

4D. What are your views on the other options, or are there any other options that the Agency should consider?

4E. Do you agree that the national park authority should have an active role in transport and traffic management?

Are there other issues that the Countryside Agency should consider?

4F. Do you agree with the role outlined for the national park authority on tourism?

Are there other issues that the Countryside Agency should consider?

4G. Do you agree with the role outlined for the national park authority on education and interpretation?

Are there other issues that the Countryside Agency should consider?

Issue 5. A national park management plan and delivery by the national park authority and others

5A. What are your views on the role of the national park management plan, and the role of a South Downs National Park Authority in co-ordinating and monitoring action by others?

Are there any other issues that the Countryside Agency should consider?

Issue 6. Working in partnership

6A. What are your views on involving other parties through joint working in order to support the work of a South Downs National Park Authority?

Are there any other options that the Countryside Agency should consider?

Issue 7. Involving local people

7A. What are your views on how a South Downs National Park Authority might involve local people?

Are there any other issues that the Countryside Agency should consider?

Part B: The draft boundary
6   Criteria and policies for national park boundaries

National parks are designated under the National Parks and Access to the Countryside Act 1949 (amended by the Countryside Act 1968 and the Environment Act 1995).

The South Downs National Park boundary must meet the **criteria** for national parks set out in the 1949 Act, as interpreted by Countryside Agency **policy**.

It will also be drawn according to the Countryside Agency's agreed **approach** to defining national park boundaries. These are set out below.

The statutory designation criteria

The 1949 Act states, in relation to national parks, that:

"5(1) The provisions of this Part of the Act shall have the effect for the purpose of preserving and enhancing the natural beauty of the areas specified in the next following subsection, and for the purpose of promoting their enjoyment by the public.

5(2) The said areas are those extensive tracts of country in England and Wales as to which it appears to the Commission that by reason of:-

a) their natural beauty, and

b) the opportunities they afford for open-air recreation, having regard both to their character and to their position in relation to centres of population, it is especially desirable that the necessary measures shall be taken for the purposes mentioned in the last foregoing subsection."

In February 2000 the Countryside Agency agreed a **policy** for applying the criteria given in section 5(2) in the light of modern society's requirements. For areas already designated as AONBs, and so already meeting the natural beauty criteria, suitability for inclusion in a national park hinges on two questions:

Is it an extensive tract of country providing, or capable of providing, sufficient opportunities for open-air recreation?

- The Agency considers that an area needs to have characteristics that mark it out as different from the bulk of 'normal countryside'; so it needs more than simply a network of rights of way. It should contain qualities that merit investment to deliver a markedly superior recreational experience. While the countryside need not be rugged or open, a sense of relative wilderness is important.

Is it especially desirable to provide for the leadership of a national park authority, with the powers and duties laid down in the Environment Act 1995?

- The Agency considers that the designation must lead to the integrated management of the area and in particular provision of a markedly superior recreational experience - more than can be achieved by local authorities alone. The special qualities that make this recreational experience must be available, promoted and interpreted to the "socially excluded", as well as to the more mobile in society, as a result of the work of that special authority.

When addressing these questions, the position of National Parks in **relation to centres of population** is especially important. People in major conurbations should have reasonable access to outstanding countryside, properly managed and promoted by a special purpose authority. In modern society, many more people seek a wide range of recreational pursuits close to where they live in a variety of landscape types. Such areas must also be capable of being reached by public transport.

For the areas not already designated an AONB, the Agency must also consider in detail if they are of outstanding natural beauty.

Defining national park boundaries

The Agency also has an agreed **approach** for defining the precise boundary of a national park, once it is satisfied that an area meets the statutory criteria set out above. These are set out in the table below.

Table 1

The Countryside Agency's approach to defining national park boundaries

1. **The Countryside Agency shall first determine in broad terms that an area of land meets the statutory criteria for designation.**

2. **It shall then in drawing a national park boundary take account together of the following considerations.**

a. Areas of high landscape quality should be included within the area of land identified for designation.

b. Areas to be included may be of differing landscape character: quality will be the key determinant rather than uniformity.

c. Areas which provide or are capable of providing a markedly superior recreational experience should be included.

d. Boundaries should include land and settlements which contribute to the rural economy and community life within the park and to the park's special qualities and purposes. Such areas should however be excluded where activities there, in particular urban or industrial development, conflict with or outweigh the essential values of the park.

e. Wherever possible, an easily distinguishable physical boundary should be chosen.

f. Where local government boundaries follow suitable lines, it may be administratively convenient to adopt them. In the majority of cases, however, they will be unsuitable.

g. Towns or villages should not normally be cut in two by a national park boundary: inclusion or exclusion should normally depend on their contribution as a whole to the character and purposes of the park.

h. Unsightly development on the edge of a national park should generally be excluded, although the possibility of its modification or screening should not be overlooked where the immediately surrounding country claims inclusion.

i. Land allocated in adopted development plans as to be worked for the quarrying and mining of important deposits on the margins of a national park should normally be excluded from the park, unless the land will be restored to a land use quality which contributes to park purposes. This approach will also apply to major industrial and commercial developments for which land is allocated in adopted development plans at the time of designation.

j. Features of scientific, historic or architectural value (eg Nature Reserves, important archaeological sites and Ancient Monuments) which are situated on the margins of a national park should be included where practicable.

3. **The statutory criteria point to the inclusion of land where both high landscape quality and markedly superior recreational opportunity exist. Not all land within the park must necessarily satisfy both criteria (a) and (c), but there should be a high degree of concurrence.**

4. **The boundary should not be regarded as a sharp barrier between areas of differing quality. In most situations there will be a transition of landscape quality and recreational experience across a sweep of land: the boundary chosen should be an easily identifiable feature within this transition.**

Table 1 explanatory notes

1. The criteria are defined in S5(2) in the 1949 National Parks and Access to the Countryside Act and shall be applied according to the Agency's policy adopted in February 2000.

2a. Landscape quality includes visual and intangible features and values. It embraces natural beauty, wildlife and cultural heritage.
    It is interpreted as the extent to which the landscape demonstrates the presence of key characteristics and the absence of atypical or incongruous ones, and by its state of repair and integrity. This is in line with the Countryside Agency's approach to landscape assessment.

2b. A variety of landscape character is an important factor in the overall amenity of the park. Usually however there will be some unifying factors, such as land use, ecosystems, historical or cultural links which bring differing character areas together to be included in a national park.

2c. Recreation means quiet countryside recreation related to the character of the area: that which allows people to enjoy and understand the special qualities of the park, without damaging it or conflicting with its purposes or spoiling the enjoyment of it by others. This definition can encompass a number of different activities.

2e. This is both for administrative reasons and for the convenience of the visiting public. Roads and railways frequently provide such a boundary.

2f. Local government boundaries are usually unsuitable because they follow no defined physical feature, may be subject to alteration and seldom conform to the limits of landscape quality or recreational value.

2g. This may include a contribution to the park's economy and community life, and a value for visitors; eg. provision of accommodation, access to public transport, information or other services.

3. This approach is in line with the Agency's policy for designation, adopted in February 2000.

7 Identifying the draft boundary for a South Downs National Park

In August 2000, the Countryside Agency began the first stage of the boundary-setting process: identifying an area of land that in broad terms meets the statutory criteria for designation.
It commissioned consultants, Landscape Design Associates (LDA), to undertake a technical study to help with this.

The resulting 'area of search' was agreed by the Agency in March 2001, and included the majority of the East Hampshire and Sussex Downs AONBs, and some areas beyond. The Agency made details of the area public, and a wide range of organisations gave the Agency their comments about it and about issues to be considered when defining a boundary.

The area of search formed the starting point for the second stage of the boundary-setting process, which was to define in detail a precise boundary on the ground. This was done by undertaking a thorough analysis of the area, considering technical advice and information from organisations and individuals, and carrying out

detailed fieldwork. This stage was again informed by a technical study carried out by LDA over the summer of 2001. The Countryside Agency then agreed a draft boundary in September 2001.

The draft South Downs National Park boundary that results from these two stages is set out in Chapter 8. The draft boundary provides:

- a diverse range of high-quality landscape character areas;
- superior recreational opportunities, which will meet the needs of a wide range of people;
- varied but unified landscapes which, when considered as a whole, make up the unique character of the South Downs;
- a national park close to urban areas and transport networks that help to ensure equal access to all;
- a landscape with the capacity and variety to accommodate predicted trends for recreation and development.

Throughout the process, the statutory designation criteria (set out in Chapter 6) have been the key factor in determining the boundary. If an area met the statutory criteria, then the detailed guidelines of the Agency's approach to defining national park boundaries were used to define a line. To aid decision making when considering how these criteria and guidelines might be applied specifically in the South Downs, further principles were defined. This chapter explains the methodology, and sets out how a draft boundary was identified.

How we arrived at a South Downs National Park draft boundary

When defining the line of the draft boundary, decisions were informed by:

- **the statutory criteria,** as set out in the 1949 Act (these are the only legally valid criteria by which a national park can be designated), as interpreted by **the Agency's policy;**
- **the Countryside Agency's approach** to the definition of national park boundaries;
- the approved area of search;
- nationally recognised methodologies for assessing the landscape in terms of its character, quality and recreation opportunities;
- professional and technical advice from the consultants, Landscape Design Associates, and from the boundary technical advisory group, local authority planning officers and other bodies, and from submissions from individuals and local groups;
- development of a transparent and consistent methodology for applying the statutory criteria and the Agency's approach to defining national park boundaries specifically in the South Downs.

Landscape character, quality and variety

The 1949 Act refers to the importance of having regard to the character of the landscape. Landscape character assessments have revealed the complexity and variety of landscape character that exists within the South Downs.

According to the Agency's approach to identifying national park boundaries, (see 2b, Table 1):

"Areas to be included may be of differing landscape character: quality will be the key determinant rather than uniformity.

A variety of landscape character is an important factor in the overall amenity of the park. Usually however there will be some unifying factors, such as land use, ecosystems, historical or cultural links which bring differing character areas together to be included in a national park."

As part of the work to define an area of search, the Countryside Agency used a landscape character area map of the South Downs, which identified ten broad landscape types. Each of these was tested for its natural beauty and opportunities for open-air recreation to see if it merited inclusion in a national park.

The character areas analysed are shown in Map A (inside the front cover of this document) and are listed in the box opposite. Four areas make up the chalk escarpment that runs from Beachy Head to Winchester. The other five are closely associated with and linked to the chalk ridge, through geographical, visual and cultural associations (Pevensey Levels, the tenth landscape character area, was excluded from the area of search). The chalk and the lowland landscapes are also linked by the river valleys that cut through them, such as the Itchen and the Arun.

Landscape character areas

Chalk landscapes

1. Eastern open chalk (including river valleys such as the Cuckmere, Ouse and Adur)
2. Central wooded chalk uplands (including the Arun river valley)
3. Western chalk uplands (including the Meon and Itchen river valleys)
4. East Hampshire Downs

Lowland landscapes

5. Hampshire Hangers
6. Wealden Greensand (including the Rother river valley)
7. Low Weald (including the Arun, Adur, Ouse and Cuckmere river valleys)
8. Forest of Bere (including the Meon river valley)
9. Coastal lowlands (including the Adur, Arun, Ouse and Cuckmere river valleys)
10. Pevensey Levels

These landscape character areas are shown on Map A.

The South Downs is not simply a chalk landscape. The draft boundary includes all the different landscape types that sum up the essence of the Downs, and which together represent a classic English lowland landscape. The presence of associated lowland landscapes adds to the distinctiveness and uniqueness of the area as a whole.

The Agency believes that all of the landscape areas, which together form the draft boundary, meet the statutory criteria for inclusion within the national park. We also believe that it will be especially desirable to include both the lowland and chalk landscapes in order to create a national park that meets our vision for the 21st century.

Testing for natural beauty

The Countryside Agency's approach to defining boundaries (see 2a, Table 1) states that: **"Areas of high landscape quality should be included within the area of land identified for designation.** Landscape quality includes visual and intangible features and values. It embraces natural beauty, wildlife and cultural heritage. It is interpreted as the extent to which the landscape demonstrates the presence of key characteristics and the absence of atypical or incongruous ones, and by its state of repair and integrity. This is in line with the Countryside Agency's approach to landscape assessment".

Standard landscape assessment techniques were used to assess natural beauty in the proposed national park area. These were based on the following tests:

- **Landscape as a resource** - the landscape should be a resource of at least national importance for reasons of rarity or representativeness.
- **Scenic quality** - it should be of high scenic quality, with pleasing patterns and combinations of landscape features, and important aesthetic or intangible factors.
- **Unspoilt character** - the landscape within the area generally should be unspoilt by large-scale visually intrusive industry, mineral extraction or other inharmonious development.
- **Sense of place** - it should have a distinctive and common character, including topographic and visual unity and a clear sense of place.
- **Conservation interests** - in addition to its scenic qualities, it should include other notable conservation interests, such as features of historical, wildlife or architectural interest.
- **Consensus** - there should be a common consensus of both professional and public opinion as to its importance, for example as reflected through writings and paintings about the landscape.

Testing for a markedly superior recreational experience

According to the Countryside Agency's approach: **"Areas which provide, or are capable of providing, a markedly superior recreational experience should be included.** This means quiet countryside recreation related to the character of the area: that which allows people to enjoy and understand the special qualities of the Park, without damaging it or conflicting with its purposes or spoiling the enjoyment of it by others. This definition can encompass a number of different activities."

Recreational experiences are influenced by, and are often dependent upon, the quality and character of the landscape; a markedly superior open-air recreational experience in the context of a national park is most likely to be achieved in a high-quality landscape. The experience must be of national importance in its quality and uniqueness, but the infrastructure must also be excellent to ensure that the experience is accessible.

The following examples illustrate the direct relationship between an exceptional landscape and a superior recreational experience.

- The expansive and dramatic setting of the open chalk landscape affords opportunities for walking, riding and cycling, as well as for more adventurous activities such as hang-gliding.

- A walk up to an escarpment hilltop offers breathtaking views over a fine unspoilt land or seascape.
- The lush and peaceful surroundings of the woodland and pasture mosaics in the Greensand Hangers, with occasional glimpses of the chalk escarpment, provide opportunities for a different type of recreational experience, in a quieter, more sheltered and verdant environment.
- For those seeking calm and undemanding recreation, there are the tranquil banks of cool, chalk-fed streams.
- The wildlife and cultural interest of the Wealden Heaths offer an opportunity to understand natural history and human influence on the countryside.

Drawing a boundary line for a South Downs National Park

Having first determined that an area of land met the statutory criteria for designation as a national park, the Agency's approach to boundary setting was then applied.

This approach is by its nature a generalised one used for the definition of any national park boundary. In identifying a boundary for the South Downs, specific principles were developed to address issues particular to the South Downs, in order to ensure that decisions were consistent and robust.

1. Development plan status

According to the Countryside Agency's approach (as at 2i), land allocated for development in an adopted development plan should be excluded. All development plans (structure plans, local plans, and minerals and waste local plans) have been reviewed to see whether there were proposed developments within the draft boundary. A number of plans in the South Downs are currently under review. Development plans that are not yet adopted were considered so that we were aware of any likely future developments. However, only policies shown in adopted plans at the time of designation were used to determine if land should be excluded.

2. Settlements

The location of the South Downs in a heavily populated region means that, inevitably, there are a number of settlements on the edge of the draft boundary. The inclusion or exclusion of each of these was assessed individually, based on the contribution they would make to national park purposes and whether they are closely linked to the Downs in character (as outlined in the Agency's approach, guidelines 2d and 2g). Villages were assessed in terms of their distinctive and/or historic character and their integration with the surrounding landscape. Towns on the edge of the boundary were individually assessed according to:

- the integrity of the historic core which contributes to the South Downs identity;
- the scale, visual impact, location and type of adverse or intrusive development;
- whether the landscape surrounding the settlement meets the statutory criteria;
- the relationship (visual, historical, socio-economic) of the town to the Downs;
- the contribution of the settlement to the purposes of the national park;
- whether the settlement offers strong recreational connections to the Downs;
- the contribution the settlement would make to the economy and community life of the national park.

Where the settlement itself did not merit inclusion, a particular approach was used to identify a boundary in urban fringe areas. In these areas, the following aspects were considered:
- Does the landscape meet the statutory criteria for designation? If so,
- Does the landscape read as part of the wider Downs landscape?
- To what extent does the urban edge impact on the surrounding landscape?
- Are there opportunities to mitigate adverse or intrusive built-up areas where the landscape itself merits inclusion?
- Is the land associated with uses that are more urban than rural in character?

3. Unifying factors and transitional landscapes

In assessing different landscape types it is important that each has strong unifying links to the essence of the South Downs (as per Table 1, 2b and 4). Strong visual links, geological associations and historical links, especially relating to land management, have been taken as unifying factors. Where the landscape is undergoing transition, in terms of both its character and quality (such that pockets of land which do not meet the criteria become increasingly frequent), a judgement has been made - in line with guideline 4 of the approach - as to when these pockets of land undermine the quality of the area, and have lost their unifying links to the character of the Downs, to such an extent that they cannot be justifiably included.

4. Fragmentation of the landscape

In a number of places the landscape has become fragmented by transport corridors and roads such as the A27 or M3. The following questions have been used to help apply guidelines 2a and 2h in particular:

- Is it a significant sweep of land or just a small 'left over' parcel?
- Does the area of land visually read as part of the wider Downs landscape?
- Is the land used for an activity more closely associated with the urban area or with the countryside?
- How accessible is the land and to what extent is it well connected to the Downs?
- How strongly does the road/ railway impact on the landscape?

5. Landscapes without public access

Areas for which there is no, or very limited, public access have been excluded from the national park on the basis that they do not meet the recreation criterion (and guideline 2b).

6. The existing AONB boundary

Where broad areas of land beyond the existing AONB boundary meet the statutory criteria for designation, they have been included. Likewise, where parcels of land within the existing AONB do not meet the criteria they have been excluded (for example where they have been developed). Otherwise, the existing AONB boundary has been used as the draft national park boundary, with changes made, where justified, according to the statutory criteria. On occasion, where the AONB boundary is now indistinguishable on the ground, a judgement has had to be made on where to draw the draft boundary line.

8 The draft boundary in detail

The draft boundary and the reasoning behind it

SECTION A: Margins of Winchester and Itchen Valley (Maps 1, 2)

Boundary section	Natural beauty	Recreation	Key considerations
The draft boundary section runs south along the A335, Bambridge Road, then west until it meets the railway. Here it continues north along the railway and edge of built development and playing fields to Wharf Bridge in Winchester. From Wharf Bridge the boundary goes south, down the Itchen Way as far as Tun Bridge. Here it heads east along Bull Drove road to the roundabout, where it joins the M3 going north, before extending back into the Itchen Valley and south to include the Winnall Moors Nature Reserve. Again, the boundary within the valley follows the edge of development and playing fields. It then follows the A33 at Abbots Barton as its western boundary before adopting the disused railway line north of Abbots Worthy.	**Landscape** • Itchen River Valley character - a typical chalk stream regarded as one of the finest in the world. • Unspoilt, intact and tranquil. Strong links to dominant chalk valley sides and memorable views. • First view of national park for visitors arriving at Winchester gateway'. • Outstanding views of Winchester water meadows from St Catherine's Hill. • Valley floor consists of a complex and intimate mosaic of fen, species-rich meadows and improved meadows and areas of distinctive parkland, e.g. Brambridge Park. • Includes areas of outstanding and classic chalk landscape which form valley sides of Itchen, e.g. around Twyford. **Nature conservation/geology** • River Itchen is designated an SSSI. Winnall Moors Nature Reserve (Hampshire Wildlife Trust) contains mosaic of grassland types and wetland areas and occupies most of Itchen Valley north of Winchester. • Includes area of Itchen Valley identified as candidate SAC between Shawford Road and Brambridge House, and between Twyford Lodge and Central Winchester (important for water crowfoot, southern damselfly, bullhead etc). • One of the last rivers in Southern England to support viable population of native freshwater crayfish. Also supports protected species such as otters. • Valley floor contains the largest assemblage of species-rich neutral grassland in England. • Includes Magdalen Hill Down Site of Nature Conservation Importance (SNCI) and butterfly reserve. **Cultural heritage** • St Catherine's Hill. • Historic water meadows of Winchester. • Historic Parks and Gardens - Brambridge, Shawford and Worthy (county register).	• Excellent public access and rights of way network. • River Valley includes the Itchen Way, and King's Way long distance footpaths, and is crossed by the Monarch's Way. • St Catherine's Hill (Iron Age hill fort) has dramatic views across the Itchen Valley watermeadows and Winchester, as well as to the wider Downs. • Shawford Park and House. • Winnall Moors Nature Reserve (tranquil and enclosed wetland landscape with attractive walks and interpretation of the area's flora and fauna). • Excellent links between sustainable modes of transport from Winchester to national park. • Key activities on and along the river include game fishing and walking, in a beautiful and tranquil setting. • Directly accessible from historic centre of Winchester. • Proposals for viaduct which crosses Itchen south of Winchester to be restored and used as footpath/cycle route. • Start/end of South Downs Way long distance footpath. • At Magdalen Hill Down Nature Reserve there is interpretation of the chalk grassland landscape.	• **Impact of M3 on recreational experience** Some noise and visual intrusion from the M3 but regarded as localised and mitigated by cutting, vegetation and old viaduct. Does not outweigh the significant benefits of the Itchen Valley north and south of the road where it crosses the valley. Access under the M3 remains good and does not fragment the recreational landscape. • **The splitting of nature conservation designations** The SSSI designation associated with the River Itchen runs for many miles south through landscapes that are not regarded as meeting the natural beauty criteria. A logical break point has been identified where the designation is narrow (i.e. along the river only) and where only land that meets the criteria is included. • **Development** Planning permission given for 'park and ride' site (600 spaces) off Garnier Road, therefore this has been excluded. Winchester North Major Development Area (MDA) potentially includes land to the east of the A272, and north of Abbots Worthy. The majority of the MDA is unlikely to be affected by the draft national park boundary. • **Existing settlements reviewed** Twyford, Northfields and Abbots Worthy are included. Twyford has a strong vernacular historic core and church landmark and forms an attractive node at the interface between the Itchen Valley and the wider chalk downs. High-quality landscape surrounds the town. Northfields is not distinctive and more recent in character. However, it is relatively small and high-quality landscape surrounds and washes over the settlement. Abbots Worthy is regarded as meeting the criteria because of its historic core vernacular architecture and association with Worthy Park. Winchester is excluded from the proposed national park due to the proportion of recent development around the historic core. The Itchen Valley, however, which enters into the heart of the city, is regarded as meeting the statutory criteria. Taking the national park into the heart of the city along the Itchen Valley enables easy access to the national park, assisting Winchester in its important role as an historic gateway. • **Playing fields** The draft national park boundary does not include playing fields along the Itchen Valley floor in Winchester. These areas are regarded as more closely associated with the urban character of Winchester rather than the wider countryside.

SECTION B: North Itchen Valley (Maps 2, 3, 4)

Boundary section	Natural beauty	Recreation	Key considerations
Through this section the draft boundary follows the line of a disused railway (marked on the ground by a strong line of vegetation) north of the Itchen from the M3 to the B3047 at New Alresford. It then follows the A31 and a road south to North End.	**Landscape** • Distinctive Itchen Valley landscape character. • Strong sense of place, with integrity and unspoilt landscape. • Memorable views to wider chalk landscape and into the valley from surrounding elevated valley sides. • High-quality tranquil flood meadows and fine enclosed landscapes, often with a parkland influence. **Nature conservation/geology** • SSSI designation, candidate SAC along River Itchen. • Valley floodplain and meadows are important for nature conservation. • One of the last rivers in Southern England to support viable population of native freshwater crayfish. Also supports protected species such as otters. • Valley floor contains the largest assemblage of species-rich neutral grassland in England. **Cultural heritage** • String of attractive historic villages occur along the Itchen Valley sides (most have Conservation Areas). • Includes remaining parkland associated with Tichborne Park and other parklands to the south of the river, including Ovington House and Avington Park. • Includes the Norman church at Easton.	• Contains the Itchen Way and King's Way long distance footpaths. • Contains numerous historic houses and associated parkland, e.g. Avington Park, which is a registered Park and Garden and country park. • Contains numerous small, attractive villages, many of which offer accommodation. • Easy access from Alresford, including access via the Watercress Line steam railway from Alton. • High density of footpaths both within the valley and to the wider chalk upland to the south. • River valley offers quiet, intimate and secluded landscapes to explore. • Key activities include walking and fishing.	• **Settlements** New Alresford is excluded as it does not meet the settlement criteria test. It does, however, offer excellent facilities on the edge of the national park and good access to it. • **Tichborne Park** Historical records revealed that the parkland associated with Tichborne House extended eastwards beyond the road south to Cheriton. In the field, however, the land to the east of the road is not regarded as meeting natural beauty criteria and the parkland character has become fragmented. Only the parkland to the west is therefore included. • **Avoid splitting settlements** The villages along the Itchen Valley are included in their entirety. It is important to identify a boundary that avoids splitting them and yet excludes land to the north of the river valley that does not meet the designation criteria.

SECTION C: North End to Upper Farringdon (Maps 4, 5, 6, 7)

Boundary section	Natural beauty	Recreation	Key considerations
The draft boundary follows Badshear and Cheriton Lane before adopting field boundaries and edges of woodland. It mainly follows the AONB boundary except for a minor addition at Charlwood Farm. At Newtonwood Farm the boundary extends north up Headmore Lane, then east along field edges to Kitcombe Lane. It then follows the built-up edge of Lower Farringdon, excluding the settlement, before following the A32 to Chawton. Here it follows the road and Wolf's Lane before continuing along Caker Stream southwards. It skirts round the built-up edge of Upper Farringdon including the settlement and then follows Hall Lane east.	**Landscape** • East Hampshire Downs Landscape Character Area. • Quiet, relatively unpopulated landscape - tranquil and rural. • Cohesive and unspoilt upland character comprising farmland, woodland and parkland. • Memorable views across classic chalk landscape. **Nature conservation/geology** • Includes Sites of Importance for Nature Conservation - Cheriton Wood, Bramdean Common, Stonybrow Plantation, Winchester Wood, Plash Wood as well as Peck Copse and Noar Copse associated with Chawton Park. **Cultural heritage** • An ancient landscape of wood pastures, remnant commons and field systems. • Woods often have signs of ancient origins - bluebells and coppice stools, e.g. Lords Wood. • Includes the historic village of Upper Farringdon, which is predominately covered by a Conservation Area. • Includes Hinton Ampner Park, Basing Park, Rotherfield Park and Chawton Park. The latter two parks are listed as Historic Parks and Gardens, and Chawton Park is associated with Jane Austen.	• The broad scale and rural character of this landscape offers a tranquil environment for quiet recreation. • Good rights of way network and quiet lanes. • Contains the Hampshire Cycle Route and Wayfarers' Walk long distance footpath. • Attractive parkland landscapes open to the public, e.g. Hinton Ampner and Chawton. • Includes Jane Austen's house which is situated in Chawton and open to the public. • Easy access to the national park from Alton and Alresford (two stations on the Watercress Line steam railway).	• **Clearly definable boundary** In some places the AONB boundary does not always follow an easily distinguishable physical boundary, e.g at Charlwood Farm. Here, the draft boundary adopts physical features on the ground, ensuring that additional areas of land, included within the boundary, meet the national park statutory criteria. • **Settlements** Lower and Upper Farringdon are both outside the AONB boundary. Upper Farringdon is designated a Conservation Area and has many attractive, distinctive and local vernacular buildings, and a strong association with its landscape setting. It is therefore included. Lower Farringdon, on the other hand, is associated with more recent development that detracts from its overall character and quality. It is therefore excluded. Chawton Village is included in the proposed national park because of its distinctive unspoilt character and association with Chawton House and Park. West of Chawton, towards Four Marks and West Tisted, the landscape quality becomes more fragmented and does not meet the natural beauty criteria.

SECTION D: The Northern Hangers (Maps 7, 8, 9, 10)

Boundary section	Natural beauty	Recreation	Key considerations
The draft boundary follows lane, field and woodland edges, including the village of East Worldham. It then follows Wyck Lane to Wyck Farm, then field edges to Binsted. North of Binsted it follows Isington Road to the railway line and then eastwards as far as Bentley Station. It then follows field boundaries across Blacknest Road and picks up the dismantled railway to Kingsley (excluded). From here, it continues along the edge of Shortheath Common (included), excluding Oakhanger and then along the road south towards Blackmoor. It skirts the edge of the hangers, following field boundaries and woodland, to also include Blackmoor Estate.	**Landscape** • Predominately the Hampshire Hangers landscape, although it does contain some areas of the Rother Valley which form a setting to the Hangers. • Continuation of the unique and distinctive chalk and greensand escarpment. • Comprises a strong sense of enclosure, seclusion and shade. • Memorable views from the escarpment across the Wealden Greensand landscape and Rother Valley. • Strong sense of place and unspoilt character. **Nature conservation/geology** • Beech and mixed woodland associated with the steep slopes of the hangers, designated SSSIs and candidate SAC. Includes Upper Greensand Hangers: Wyck to Wheatley. • Many other hanger woodlands are designated Sites of Importance for Nature Conservation, including Hartley Wood and Warner's Wood. • Includes Shortheath Common, an SSSI and possible SAC which contains a mosaic of mire and bog as well as heathland, and is also a Local Nature Reserve. **Cultural heritage** • Includes the historic villages of East Worldham, Wyck and Binsted which are associated with the Hangers. • Includes Hartley Mauditt church and deserted medieval settlement earthworks (a Scheduled Ancient Monument). • East Woldham contains an attractive church that acts as a landmark, seen on the skyline when ascending the Hangers.	• Includes the whole of the Hangers landscape and The Hangers Way long distance footpath, enabling integrated recreational management of this important landscape. • Area contains a significant number of public rights of way providing a high-density network. • Includes Bentley Station, which offers immediate access into the national park via train from London and the north. • Includes Shortheath Common Local Nature Reserve, which offers open access and interpretation. • Includes the Blackmoor Estate and settlement, which produces apple juice and associated products and sells them at its farm shop. • Intimate, enclosed landscape both in the Hangers and areas of heathland, which contrasts with the expansive views gained from the edge of the greensand ridge across the Rother Valley. • Includes attractive rural settlements along Hangers and associated churches.	• **Exclusion of Alice Holt Forest** Alice Holt Forest is owned and managed by the Forestry Commission. It is open to public access and offers some excellent opportunities for open-air recreation. However, the areas of woodland are fragmented by roads and recent development, which detracts from the overall quality of the landscape. While the woodland areas to the west are more closely associated with the Hangers landscape, and thus the South Downs, the remaining areas of woodland are not. On balance it is regarded as preferable to exclude the whole of the Forest, thus avoiding having to split its management. • **Access from rural station of Bentley** Bentley Station lies at the edge of the proposed national park. The landscape to the south of the station is regarded as meeting the natural beauty criteria, and footpath access from the station to the wider proposed national park is regarded as providing a highly desirable recreational link. • **Exclusion of mineral extraction sites** Lode Farm, Kingsley and Selbourne Brickworks are excluded from the proposed national park because they are currently being worked or have been allocated in the Hampshire, Portsmouth and Southampton Minerals and Waste Local Plan (1998), and occur on the edge of the draft national park boundary. • **Inclusion of Blackmoor Settlement** Blackmoor Village is regarded as meeting the criteria because of its attractive buildings and strong estate character, and landscape setting, but also because of the recreational opportunities presented by its apple pressing and shop.

SECTION E: Upper Rother Valley and Weavers Down (Maps 9, 10, 11)

Boundary section	Natural beauty	Recreation	Key considerations
From Blackmoor the draft boundary follows Blackmoor Road and then Petersfield Road south. At the junction with the A325, the boundary goes west along Benhams Lane. It then goes south and follows the built edge of Greatham (excluding the village) and then field boundaries and the edge of woodland to Longmoor Road. It then continues east along the edge of the Longmoor Camp (a Ministry of Defence training village) and airstrip, excluding these areas. It then extends eastwards along a bridleway to Griggs Green and then along the built edge of Liphook. It then adopts Portsmouth Road south and then the railway line north and then the southern built edge of Liphook, before joining Highfield Lane, north.	**Landscape** • Wealden Greensand character area, within which the Rother Valley forms a significant component. • Memorable and distinctive views to the Hampshire Hangers to the west and the chalk escarpment to the south. • Diverse landscape of river valley character, heathlands, acidic woodlands and farmland. **Nature conservation/geology** • Includes part of the Woolmer Forest SSSI and SPA designation (Weavers Down and Holly Hills). • Includes Chapel Common SSSI, Rother Valley Riverside Walk Local Nature Reserve and Wheatsheaf Common SNCI. • Includes the River Rother, important for protected species such as the otter and three red list bird species. **Cultural heritage** • Includes Hollycombe and Shufflesheep Historic Park and Garden. • Includes the historic (Norman) settlement of Petersfield (population 12,000) which has strong historical links to the South Downs landscape, and grew as a result of the wool and cloth industry and trade. In the 17th century it became a coaching town for those travelling from London to the South Coast. Its historic core is designated a Conservation Area. • Includes the historic churches of Greatham, and West Liss. • Includes the Victorian railway settlement of Liss, which has strong historical and military connections with Longmoor Camp and Woolmer Forest.	• Includes the Longmoor Inclosure, owned by the Ministry of Defence. This area is generally open to public access and only occasionally subject to short-notice closure for military use. This landscape provides an excellent lowland heath experience and forms the largest area of heath within the proposed national park. It has an estimated 30-40,000 visitors each year. • Includes the Rother Valley Riverside Walk Local Nature Reserve from Liss to Longmoor, which is accessible to walkers, cyclists and horse riders, as well as wheelchair users. • Includes the numerous public rights of way that extend out from Petersfield and Liss to the wider landscape, including the southern end of the Hangers Way which connects the Hangers with the chalk escarpment. • Includes the Sussex Border Path connecting Haslemere with the chalk escarpment and the South Downs Way. • Includes Petersfield, which offers many tourist facilities such as tourist information, museum, cycle hire and integrated access to the national park via rail and bus. Petersfield offers recreational opportunities such as two golf courses with public access, memorable views to the Downs and Hangers, as well as Wheatsheaf Common and Heath Pond where boating is popular. • There are approximately 20 footpath and road crossings over the A3 between Petersfield and the Longmoor enclosure. This ensures continuity of the recreational landscape despite the presence of the transport route.	• **MoD land** Land in MoD ownership includes the Woolmer Forest and Longmoor Inclosure. Heathland within MoD ownership to the north of the A3, while meeting the natural beauty criteria, is excluded on the basis that it is fenced and regularly closed for use as a military training ground and therefore does not offer markedly superior recreational opportunities. However, the land under MoD ownership to the south of the A3 (Longmoor Inclosure and Weavers Down) is of high quality and is regularly open to the public for recreational use. This southern area therefore meets the criteria for inclusion. • **Settlements** The Upper Rother Valley was considered in detail to assess whether it met the designation criteria and to determine a suitable boundary on the ground. The settlements of Petersfield and Liss were reviewed as part of this process. Petersfield is regarded as meeting the criteria for inclusion within the proposed national park because of its high-quality townscape and strong relationship with its outstanding landscape setting. While the case for Liss was borderline, it is exceptionally difficult to determine a clearly identifiable and strongly defendable boundary on the ground which excludes Liss. In light of the Longmoor area meeting the criteria for designation, it is concluded that the whole of the Upper Rother Valley be included. Liphook does not meet the criteria due to recent development and weak links to the surrounding landscape. This results in the exclusion of high quality areas north of Liphook, such as Bramshott Common. • **The A3 corridor** The A3 and rail line fragment the landscape and recreational experience of this area to some degree. However there are still a significant number of footpath and road bridges that provide continuity of access. The visual impact of the road is considered to be very localised, as is its noise intrusion, as for much of the route the road is in a cutting. On balance, the A3 and rail corridor are not regarded as having a significant detrimental impact on the recreational experience of the landscape and do not justify the exclusion of this area.

SECTION F: South of Haslemere (Maps 11, 12)

Boundary section	Natural beauty	Recreation	Key considerations

The draft boundary follows the county boundary, north of the railway line to Hammer Bottom, where it follows the edge of built development around Hammer and Camelsdale, excluding them.
 	
It then follows a bridleway, stream and the boundary of the Black Down National Trust Estate to the Petworth Road, where it continues east as far as the junction with the A283.
	Landscape
	• Comprises the Wealden Greensand and Low Weald character areas.
	• Distinctive and remarkably unspoilt landscape of woodland, heath and farmland.
	• The densely wooded greensand ridges provide topographic variety and visual interest.
	
	Nature conservation/geology
	• Includes Linchmere Common and Marley Lane Local Nature Reserves.
	• Includes SNCIs associated with woodland and common, e.g. Jay's Copse and Black Down.
	
	Cultural heritage
	• Contains numerous remote and historic settlements set within an unspoilt landscape setting, e.g. Kingsley Green and Fernhurst.
		• Includes the National Trust estates of Marley Common and Black Down.
		• Includes the Sussex Border long distance footpath.
		• Contains a significant number of public rights of way.
		• Opportunities to manage recreation in area to retain the deeply rural and remote characteristics of the area.
		• Associated with the greensand ridges are areas of woodland and sunken lanes, which provide a mysterious and remote recreational experience.
		• Contains significant areas of landscape that provide tranquil surroundings in which to walk, ride and cycle.
		• This landscape contains the highest percentage of registered common land within the proposed national park, potentially offering many areas of open access.
			• **Settlements**
			  Currently the settlements of Camelsdale and Hammer are split by the AONB boundary. The draft national park boundary excludes these settlements on the basis that they have been developed and do not meet national park criteria.
			
			• **National Trust Estate**
			  The boundary does not extend into Surrey and in this area the administrative boundaries have been adopted. As a result, the Black Down National Trust estate is split by the draft national park boundary.

SECTION G: North of Petworth (Maps 12, 13, 14)

Boundary section	Natural beauty	Recreation	Key considerations
After going south down the A283, the draft boundary follows a minor road to Shillinglee Park. It then follows a track to Park Mill Farm and then to Piper's Cottages before joining Piper's Lane to Balls Cross (included). From here it follows a track east and then south to Medhone Copse. It then continues along the edges of woodland and field boundaries to Idehurst Copse, where it picks up the River Kird.	**Landscape** • Low Weald landscape character area. • Densely wooded character with a patchwork of small fields. • Includes the more undulating landscape to the east of Northchapel which has a close visual association with the greensand ridges to the west. • Relatively unpopulated and deeply rural landscape which remains unspoilt. **Nature conservation/geology** • Includes Copygrove Copse and Frith Wood SNCIs which are ancient woodlands. • Includes Ebernoe Common SSSI, candidate SAC and National Nature Reserve, which comprises ancient woodland. • Includes Shillinglee Lake SSSI. • Includes Piper's Kiln and Mercers Copses, Colhook Common and Bittles Field SNCIs. • Includes Idehurst Copse, designated a candidate SAC, a component of The Mens SSSI, designated for size, structural diversity and rich fungal and lichen floras. **Cultural heritage** • Contains the historic village of Northchapel. • Contains the ancient wooded landscape of the Low Weald, which shows signs of past coppice management. • Includes Petworth House and Historic Park and Garden.	• Includes an intimate and remote wooded landscape suitable for those who wish to explore a more tranquil environment. • Medium density of public rights of way throughout area. • Fishing available at Northchapel. • Extensive areas of common land and woodland to explore, with potential for open access. • Petworth House is a 17th century palace set in stunning parkland designed by 'Capability Brown' and immortalised in paintings by Turner.	• **Landscapes in transition** This landscape includes a significant area of ancient and semi-natural woodland, much of which is designated as SNCIs or SSSIs. Many of these woodlands are also registered as areas of common land and this, coupled with common land to the west, represents one of the highest concentrations of common land in the South Downs area. There is also an increase in the dispersed pattern of farmsteads and small villages, giving rise to a more populated character. On this basis, a line is drawn to include land that has a high concentration of designated woodland (i.e. Ebernoe Common SSSI and candidate SAC), landscapes that demonstrate cohesiveness and integrity and also have strong unifying factors which link them to the South Downs. • **Settlements** Northchapel is currently split by the AONB boundary. A review of the settlement revealed that it should be included, primarily because of its intact historic character and strong association with a high-quality landscape setting. Other settlements further east, including Plaistow, Kirdford and Wisborough Green, were also assessed. While they are all regarded as attractive, with intact historic cores, they do not reflect the typical South Downs vernacular or urban form and have stronger links to the wider Surrey landscape.

SECTION H: North West of Pulborough (Maps 14, 15)

Boundary section	Natural beauty	Recreation	Key considerations
The draft boundary heads south following the edge of Idehurst Copse, excluding Idehurst Farm, then joining the Fittleworth Road and Pallingham Lane to Pallingham Manor House, where it crosses the River Arun. From here, it picks up the Wey South Path as far as the A283, Stopham Road.	**Landscape** • Comprises the River Arun valley and Low Weald landscape character area. • Attractive views across this deeply rural and well-wooded valley landscape. **Nature conservation/geology** • Includes Park Farm Cutting (SSSI and RIGS designation) which is the only good exposure of the Sandgate Beds in the southern Weald. • Includes Wisborough Green pastures and Badlands Meadows SNCIs and The Mens SSSI and candidate SAC. • Includes Arun River SSSI. **Cultural heritage** • Includes Park Mound motte and bailey and Stopham Bridge, both impressive landmarks and scheduled ancient monuments.	• Includes part of the Wey South Path and a high density of public rights of way. • Fishing on River Arun and River Rother. • Opportunities for walking, cycling and horse riding, both in the wooded landscape to the west and more open landscape of the Arun Valley.	• **Conservation sites** A number of designated conservation sites are split or excluded by the existing AONB boundary. The draft national park boundary includes, where feasible, conservation sites that lie adjacent to it. In the case of the River Arun Valley landscape, the area towards Billingshurst is not regarded as meeting natural beauty criteria, despite the continuation of the river SSSI designation. This SSSI designation is therefore split by the draft boundary.

SECTION I: Pulborough Marshes (Maps 15, 16)

Boundary section	Natural beauty	Recreation	Key considerations
The draft boundary follows the River Arun to the south of Pulborough. It then follows the built edge of Pulborough as far as the A283. South of Wickford Bridge it cuts across the road following a track and then the edge of woodland associated with Hurston Warren. It excludes Hurston Place Farm and then uses lanes south, rejoining the A283 at Parham Airfield, and continues along the A283 as far as the junction with Clay Lane.	**Landscape** • Forms part of the wider Sussex Wealden Greensand and includes the flood plains of the Arun and Rother. • Distinctive, relatively flat, wetland landscape covering an extensive area of grazing marshes, with views to the chalk escarpment. **Nature conservation/geology** • Includes Pulborough Brooks, RSPB Nature Reserve, SSSI, Ramsar and SPA, which is outstanding for its ornithological importance. • Includes Hurston Warren SSSI, which is important for its heathland (dry and wet), quaking bog (one of the best examples in the South East) and woodland. • Includes Bog Common, which forms part of the Parham Park SSSI. **Cultural heritage** • An historic landscape containing remnants of occupation from the Romano-British period, particularly north of Wiggonholt. • Includes a number of scheduled ancient monuments – Roman station, Hardham Priory (an attractive landmark when viewed from the south) and the site of a Roman bathhouse. • Includes Parham Park and House registered Historic Park and Garden. • Includes the historic hamlet of Wiggonholt, designated a Conservation Area.	• This landscape provides opportunities for a wide range of recreational activities in a high-quality wetland/lowland landscape, with outstanding views of the chalk escarpment. • Includes the continuation of the Wey South long distance path. • Existing footpath network is low-medium density but provides reasonable access to the meadows and excellent views across to the chalk. • Includes the West Sussex County Council Heritage Trail. • Potential to improve the use of the dismantled railway from Pulborough to Midhurst. • Includes Hurston Warren golf course and associated heathland habitats, with access via public rights of way. • Excellent opportunities for bird watching. • RSPB reserve at Pulborough Brooks and Waltham Brooks, with visitor centre and interpretation at Pulborough Brooks. • Includes Parham Deer Park and House (approximately 28,300 visitors each year). • Fishing along river.	**Unsightly developments** • The corridor of the A29 and rail line, including the settlements of Coldwaltham and Watersfield, forms a very narrow strip of land either side of high-quality landscapes. Development here is not regarded as sufficiently adverse to exclude and is therefore included.

SECTION J: South of Storrington (Maps 16, 17)

Boundary section	Natural beauty	Recreation	Key considerations
The draft boundary goes south down Clay Lane and then north up the Amberley Road and continues along the southern built-up edge of Storrington. East of Chantry Lane Sandpit the boundary rejoins the A283, passes round the junction with the A24 and continues along the southern edge of the road. It then follows field boundaries, woodland and the edge of the Rock Common sandpit and Rock Farm to join the A24.	**Landscape** • Low Weald landscape character area. • Dramatic views to the chalk escarpment. • Borrows character and sense of place from the escarpment. • Unspoilt landscape which forms a high-quality setting to the chalk. **Nature conservation** • Includes SSSI sites along the escarpment, designated for important chalk grassland. **Cultural heritage** • No outstanding places of interest.	• Good access to the chalk escarpment from Storrington via country lanes and public rights of way. • Includes the South Downs Way long distance footpath. • Provides high-quality landscape in which to appreciate the scale and presence of the chalk escarpment. • Storrington town acts as a gateway into the national park, offering a wide variety of visitor facilities, including cycle hire and accommodation.	**Mineral extraction sites** These include Chantry Lane Sandpit (working), Angells Sandpit (restoration), Sandgate Park (working), Hampers Lane (working), Rock Common Sandpit (working/landfill) and Windmill and The Rough (landfill). Where relevant, restoration plans have been taken into account in defining the draft boundary. In the Horsham District Adopted Local Plan there is a policy to restore mineral extraction sites to improve amenity and access south of Storrington, along the lines of a Country Park. These proposals are long-term aspirations and are dependent upon landowners. The extraction sites are currently being worked or used for landfill (with the exception of Angells Sandpit) and are therefore excluded from the proposed national park. Angells Sandpit will be filled and restored to amenity heathland and will connect up with the Sullington Warren National Trust land to the north. However, the Angells Sandpit is only narrowly connected to the wider national park and separated by the A283. For this reason, it is also excluded. **Exclusion of Nature Conservation Sites adjacent to the boundary** Chantry Lane SSSI is designated for its important geology but is a working sandpit. It therefore lies adjacent to the proposed draft boundary but was excluded on this basis. Sullington Warren is an SSSI because of its heath and woodland habitats. It is also owned by the National Trust. Although it forms a very attractive landscape, it is separated from the wider proposed national park landscape by Angells Sandpit and the A283, and is therefore not included. **Settlements** Storrington is excluded from the proposed national park because of the scale of its more recent development and because it is separated from the proposed national park by a number of working mineral extraction sites.

SECTION K: Low Weald: Washington to Cooksbridge (Maps 17, 18, 19, 20, 21, 22)

Boundary section	Natural beauty	Recreation	Key considerations
The draft boundary follows the A24, Hole Street and Spithandle Lane before continuing along field boundaries as far as the dismantled railway, flanking the Adur floodplain. It then follows field boundaries to the built-up edge of Steyning, Bramber and Upper Beeding, excluding the settlements. It continues along field boundaries and the edge of woodland between Upper Beeding and Shaves Wood before adopting the B2117 to Hurstpierpoint. From here, it skirts the urban edge of Hurstpierpoint, Hassocks, Keymer and Ditchling and continues along field boundaries, including the village of Streat, until it meets the railway line, which it follows east to Cooksbridge.	**Landscape** • Low Weald landscape character area, and includes the River Adur Valley south of the escarpment. • Includes attractive, high-quality, cohesive and well-managed farmland, woodland blocks (particularly in the Wiston area), and attractive lanes and buildings. • Dramatic views to and from the chalk escarpment, particularly at Devil's Dyke. • Unspoilt landscape that borrows character and sense of place from the escarpment and forms a high-quality setting to the chalk. • Includes some of the most dramatic South Downs landscapes such as Devil's Dyke, Fulking Hill and Wolstonbury Hill. **Nature conservation/geology** • Includes Tottington Wood SNCI. • Includes SSSI woodland and grassland along the chalk escarpment between Fulking and east of Clayton. • Includes Ditchling Beacon Nature Reserve. **Cultural heritage** • An historic landscape depicting the former sheep management and rural economy of the area, which linked the chalk with the weald. This is revealed in the pattern of parish boundaries, drove roads and historic settlements of Streat, Fulking, Poynings, Pyecombe and Clayton. • Includes Plumpton Place Historic Park and Garden. • Includes the landmark Jack and Jill Windmills. • Includes many archaeological and Scheduled Ancient Monument sites along the escarpment, including Wolstonbury Iron Age hill fort.	• Fishing on River Adur. • Includes areas with medium and high density of public rights of way. • Includes Downs Link and Sussex Border Path long distance footpaths. • Greatest concentration of hang-gliding and paragliding off the escarpment within the proposed national park. • Includes significant National Trust estates along the escarpment. • Includes spectacular walks along the top and foot of the escarpment. • Provides a range of recreational experiences, from the gently undulating and wooded/enclosed character of the Weald to the dramatic and open views from the escarpment edge. • Visitor facilities (parking etc.) available at Devil's Dyke, and other parking and viewing points available at Summer Down and Ditchling Beacon. • Includes significant areas of landscape owned by the National Trust, offering excellent access. • Includes areas of the Low Weald landscape - a high-quality landscape offering excellent opportunities to appreciate the magnificent views of the chalk escarpment. • Potential for improved interpretation of the historic landscapes and ancient parishes.	**Settlements** A number of historic market towns are located on the edge of the draft national park boundary. The settlements of Steyning, Bramber, Upper Beeding, Hurstpierpoint, Hassocks, Keymer and Ditchling were considered. In most cases, the settlements demonstrate many attractive and positive assets, such as historic cores and views, and historic connections to the South Downs. However, they also contain significant areas of more 'ordinary' development around the historic cores, which is not in keeping with the character and quality of the South Downs. Equally, these towns are not surrounded by high-quality landscape. From a recreation perspective many of them offer visitor facilities and attractions, but none are regarded as outstanding. Their ability to act as gateways to the proposed national park is not regarded as adversely affected by their exclusion. Thus, on balance, all of the above settlements are excluded and the draft boundary adopts the often tree-lined southern built-up edges of these settlements. **Unifying factors and transitional landscapes** This section comprises the Low Weald landscape adjacent to the chalk escarpment. It is transitional both in terms of its quality and the extent to which it borrows character and sense of place from the escarpment. Close to the escarpment it demonstrates a close association with the chalk, both historically and visually. Further north, these unifying factors and association with the Downs. The draft national park boundary includes land that meets the national park statutory criteria but also demonstrates unifying factors and association with the Downs. **River Adur floodplain** The Adur floodplain, water meadows and grazing marshes to the north of Bramber, are designated as an SNCI and there are a number of footpaths across the floodplain, including the Downs Link long distance path. This landscape is very attractive in places (particularly further north), although in these areas it is less associated with the chalk escarpment. Further south, the landscape condition is mixed and on balance not regarded as meeting natural beauty criteria, particularly adjacent to Bramber and Upper Beeding. Although there is a case for including the more attractive areas further north within the proposed national park, this would create a 'hole' of quite a significant size in the proposed national park designation to the south (caused by the exclusion of lesser quality floodplain and the settlements of Steyning, Bramber and Upper Beeding). On balance, it is concluded that the Adur floodplain north of Bramber should be excluded. **Shoreham cement works** This site is located in the Adur valley to the south of Steyning and Upper Beeding and is a former chalk quarry. It is currently allocated for development (business park and residential) in the Horsham and Adur Adopted Local Plans (1997 and 1996 respectively). It is located well within the existing AONB. Whilst as a development site it would not be compatible with a national park, it would not be possible to exclude the site from the proposed national park without creating a 'hole' which is regarded as undesirable. For these reasons, and because these proposals are currently being put forward in the context of existing AONB policies, the site has been included. **Extraction site at East Chiltington** This is a proposed extension to an existing extraction site within the East Sussex and Brighton and Hove Minerals Local Plan (Adopted 1999). This is a relatively small site that is not widely visible from the surrounding landscape, and does not occur on the edge of the draft boundary. It is therefore included.

SECTION L: Lewes (Maps 22, 23)

Boundary section	Natural beauty	Recreation	Key considerations
This section of the draft boundary runs from Cooksbridge to Ringmer to the north of Lewes. It follows Hamsey Lane and the River Ouse for much of its length, before following hedgerows and a track further east. It joins the B2192 and continues to Ringmer.	**Landscape** • Comprises the Low Weald and Ouse Valley landscapes at the foot of the chalk escarpment. Significant and memorable views to the chalk. • High-quality rural landscape and attractive water meadows along the Ouse Valley, which borrow character and sense of place from the chalk. • Includes the historic settlement of Hamsey, which sits on a small knoll within the floodplain. • Variety of landscape character. • Lewes sits between and is embraced by areas of chalk down. **Nature conservation/geology** • Includes disused railway line and South Malling SNCI, which is important for rich chalk grassland and hoverflies. • Includes Offham Marshes SSSI, important for alluvial grazing marshes. **Cultural heritage** • Includes the historic Saxon market and inland port of Lewes, including its Norman castle and a superb collection of historic buildings. • Includes the historic churches of Hamsey and Malling to the north of the town. • Includes ruins of St Pancras priory in Lewes, which dates from the 11th century. • Includes the site of the battle of Lewes to the west of the town, where Henry VII was defeated by Simon de Montford in 1264. • Includes Anne of Cleves' House and Museum, situated in the town of Lewes.	• Includes the important footpath link from Lewes to Hamsey along the Ouse Valley. • Lewes is a popular tourist destination, offering many tourist attractions, including the castle, museum, and workplace attractions, as well as tourist facilities (including a tourist information centre). • The Low Weald and Ouse Valley landscapes contrast with that of the chalk escarpment and provide a variety of memorable recreational experiences. • Anne of Cleves' house illustrates life in the 17th and 18th centuries. • The castle is of historic interest and provides panoramic views of the Downs and the Ouse Valley. • The Barbican House Museum demonstrates the history of the people of Sussex. • Has literary associations with William Morris who wrote: "'You can see Lewes lying like a box of toys under a great amphitheatre of chalky hills". • Includes the Pells area - 'L' shaped backwater and the adjacent spring-fed swimming pools that are the oldest in England. • Good footpath connections from Lewes to the Downs and to water meadows to the north of the town. • Potential for restoration of Lavender Line steam railway, to connect with Lewes.	**Inclusion of Lewes** Lewes has a significant historic core, which remains remarkably intact and is designated a Conservation Area. Whilst there are also significant areas of more recent development (both housing and industrial) it is concluded that these do not significantly detract from, or disrupt, the integrity of the historic core, nor do they interrupt the strong sense of arrival at the historic town. The town still displays strong visual and historic links with the Downs landscape, the former also being true for areas of more recent development. The town offers significant tourist facilities and sustainable transport options, both from further afield (London) and also into the wider proposed national park. Its overall sense of place remains very strong and it offers a positive and memorable experience within the proposed national park. It is therefore included within the draft boundary. **Cooksbridge** This settlement comprises mainly recent development and lacks a distinctive historic core. It is therefore excluded. It does, however, contain a train station and could act as a rural gateway into the proposed national park. **The River Ouse Valley to Barcombe Mills** The Ouse Valley between the town of Lewes and Hamsey church has a strong visual link to the chalk escarpment. The historic church of Hamsey sits on a small knoll of raised land in the floodplain enclosing the Ouse water meadows to the south. This area contains a number of nature conservation sites. Although there are some views to the more recent edges of Lewes i.e. Malling and Landport, these are limited and screened by existing vegetation, and are not sufficient to detract from the area's natural beauty. This is equally true of the small-scale pylons within the floodplain, which are not visually significant. North of Hamsey to Barcombe Mills the river floodplain forms a more topographically uniform area of attractive countryside, although its condition is mixed and it has a less strong visual connection of borrowed character from the chalk escarpment compared with the area south of Hamsey. There is little woodland and there are no national nature conservation designations within the floodplain, and limited recreational opportunities (including no public rights of way). On this basis, this area of the Ouse floodplain is regarded as not meeting the statutory criteria for designation and is excluded.

SECTION M: Eastern Low Weald (Maps 23, 24, 25)

Boundary section	Natural beauty	Recreation	Key considerations
From Ringmer the draft boundary follows field boundaries and the edge of woodland, heading south to Decoy Wood. The boundary follows field boundaries and woodland though this area to Glynde Bridge, where it adopts the track to Balcombe pit (excluded) and then the A27. It continues along the southern edge of the A27 until it reaches the Gibralter area south of the road, which is excluded. The boundary then goes north of the A27 to include land at Newhouse Farm, so following the existing AONB boundary. It then goes south, rejoining the A27 and continues on its southern side eastwards to Polegate.	**Landscape** • Includes the Low Weald landscape adjacent to the chalk escarpment. • Includes the Cuckmere Valley and water meadows adjacent to Glynde. • Includes Glynde and Firle parkland landscapes. • Has a strong sense of place associated with the dramatic presence of the chalk escarpment. • The landscape is gently undulating, comprising a strong composition of chalk grassland, arable fields, woodland blocks and contrasting river flood meadows. **Nature conservation/geology** • Includes the SSSIs associated with the steep slopes of the escarpment and significant areas of chalk grassland. • Includes Tilton Wood SNCI. • Includes the SNCI west of Milton Street. **Cultural heritage** • Includes numerous historic villages at the foot of the escarpment, such as Glynde, Willmington and Alfriston. • Includes the registered Historic Parks and Gardens of Glynde Place and Firle Place. • Includes The Long Man Chalk landmark. • Includes Charleston Farm, which was the home of Virginia Woolf.	• Includes attractions such as Glynde Place, Firle Place, Drusillas Zoo and Alfriston Clergy House. • Includes the Vanguard Way and Wealdway, both of which connect to the South Downs Way long distance footpaths. • The area is well served by an extensive public rights of way network. • Includes the train station of Glynde. • Includes the attractive Charleston Farm, which is open to the public.	• **Balcombe Pit** The former quarry has been allocated for business development and has thus been excluded. • **Wilmington/Polegate and Selmeston bypasses** These bypass routes are identified in the Wealden Adopted Local Plan (1998) as safeguarded routes subject to further studies. Only the Selmeston proposed route comes south into the proposed national park. In light of this, the existing southern side of the A27 was regarded as the most suitable boundary. • **Land north of the A27** There are a number of sites of high quality north of the A27 (for example Abbot's Wood). These are not included due to the broad area not meeting the natural beauty criteria and their weak association with the Downs. The A27 was judged to be the most appropriate boundary in this area.

SECTION N: Eastbourne (Maps 25, 26, 27)

Boundary section	Natural beauty	Recreation	Key considerations
The draft boundary runs along the edge of the built-up areas of Polgate and Eastbourne.	**Landscape** • Includes the dramatic eastern escarpment of the South Downs, which is characterised by its steep topography, dense wooded areas and chalk grassland. • A stunning and memorable landscape. **Nature conservation/geology** • Includes Willingdon Downs SSSI, which is important for its species-rich grassland. • Includes Seaford to Beachy Head coastal SSSI, important for its biological and geological features, and for terrestrial as well as marine habitats. **Cultural heritage** • Includes numerous Scheduled Ancient Monuments (many of which are tumuli), which line the edge of the escarpment. • Includes Filching Manor, Polgate.	• Excellent footpath connections from the urban edge up onto the Downs. • The start/finish of the South Downs Way long distance route. • Includes three golf courses on the edge of Eastbourne, all of which have access and offer opportunities to enjoy the Downs landscape and dramatic escarpment. • Eastbourne acts as a significant gateway, with substantial facilities, bedspaces and sustainable transport.	• **Urban edge boundary** The boundary along the edge of Eastbourne was carefully assessed to ensure that the draft national park boundary ran up to the edge of the built-up area where land adjacent to the urban edge meets the statutory criteria. • **Splitting the Seaford to Beachy Head SSSI** The draft national park boundary splits the SSSI designation, which continues to run north-east along the seafront of Eastbourne. The coastal foreshore adjacent to Eastbourne is considered more urban in character and is not regarded as meeting natural beauty criteria. It is therefore excluded.

SECTION O: Coastal Foreshore (Maps 27, 28)

Boundary section	Natural beauty	Recreation	Key considerations

From the edge of Eastbourne, the proposed national park boundary follows the mean low water mark along the coast as far as the groynes at Seaford.	**Landscape**
	• Includes the Sussex Heritage Coast. Dramatic chalk cliffs, wave-cut platform and other geological features, combined with the seascape and chalk grassland of the Downs, make a dramatic and outstanding landscape.
	• Includes significant areas of chalk grassland along the cliffs.
	• One of the few remaining lengths of undeveloped coast in South East England.
	• Includes Cuckmere Valley and Cuckmere Haven; distinctive chalk river system and landscape.
	
	Nature conservation/geology
	• Includes Cuckmere Haven/Seven Sisters Local Nature Reserves.
	• Includes the Seaford to Beachy Head SSSI, which is important for its terrestrial as well as geological features.
	
	Cultural heritage
	• Includes Beachy Head lighthouse - a national landmark.
	• Includes the ancient settlement at Birling Gap, a Scheduled Ancient Monument.
		• Includes the South Downs Way long distance footpath, which runs along the chalk cliffs and offers spectacular sea and land views.
		• Includes the Vanguard Way long distance path, which passes along the attractive and distinctive Cuckmere Valley.
		• A range of visitor facilities are offered at Exceat.
		• Includes the coastal foreshore and access points to the foreshore at Cow and Birling Gaps as well as Cuckmere Haven, offering opportunities for walking, beach combing and swimming.
		• Includes the Seven Sisters Country Park, which provides public access for quiet enjoyment and also pursues conservation and agricultural objectives.
		• Includes extensive areas of National Trust estate and areas of open access.
		• Includes the countryside centre at Beachy Head.
		• Hang-gliding and gliding from Beachy Head.
			Management of the coastal area
			• The draft national park boundary extends as far as the mean low water mark. Under the 1949 Act it is not possible to extend a national park boundary beyond this mark.
			  However, it is recognised that the special quality of this area is in part due to the interplay of the spectacular white cliffs and seascape, which makes up many of the dramatic views.
			  The coastal sea area is under pressure from increased numbers of water sports and potential for windfarms, which could undermine the quiet enjoyment of this foreshore and clifftop landscape.
			  The future management of the coastal area and the need to work with other organisations will therefore be an important consideration for a national park authority.

SECTION P: Seaford to Worthing (Maps 28, 29, 30, 31, 32, 33, 34, 35, 36)

Boundary section	Natural beauty	Recreation	Key considerations
This section stretches from the edge of Seaford to Newhaven, Peacehaven, Saltdean and Rottingdean, along the northern edge of Brighton and Hove to Shoreham and Worthing. The built edge, roads (mainly the A27 bypass) and field margins are used.	**Landscape** • Comprises Eastern Open Chalk character area and the river valleys of the Ouse and Adur. • Includes land that reflects a strong, open and bold chalk landscape character. • Includes landscape which is dramatic and sculptured, with a strong sense of place, space and sky. • Includes significant areas of remote and tranquil landscape close to major populations. **Nature conservation** • Includes Local Nature Reserves such as Hollingbury, Stanmer/Coldean, Lancing Ring. • Includes numerous SNCIs designated for their chalk grassland, such as Halcombe Farm (Peacehaven). **Cultural heritage** • Includes a number of Scheduled Ancient Monuments, including Hollingbury Castle and Cissbury Ring. • Includes the Historic Park and Garden of Stanmer and Coldean. • Includes historic villages of Falmer, Telscombe and Findon.	• Includes land adjacent to urban areas, and thus areas that are easily accessed by footpath, bridleways and cycle routes. • Includes Monarchs Way, Downs Link, and Sussex Border Path long distance footpaths. • Includes significant area of landscape with common rights and open access from the urban edge. Offers opportunities for quiet recreation in landscapes which appear remarkably remote and of outstanding landscape quality, despite their close proximity to urban centres. • Includes hang-gliding and paragliding sites located north of Seaford. • There are a significant number of orienteering courses along the urban edge between Worthing and Shoreham. • Includes some outstanding golf courses with footpath access and superb views across the Downs. • Includes Stanmer Park, a popular area for picnics, walking and events in superb downland parkland setting. • Includes outstanding archaeological sites, such as Cissbury Ring, which are easily accessible via footpaths and tracks from the urban edge. • Historic villages of Falmer, Telscombe and Findon all provide visitors with a sense of early settlements, typical of the chalk.	• **Use of the AONB boundary** Where possible, the existing AONB boundary has been used in the urban fringe area north of the coastal settlements. However, in places, and especially along the new A27 bypass, it has been difficult to identify the precise AONB boundary. Here a judgement has been reached about whether small areas of land, fragmented from the wider downland, should be included if they still meet the criteria. Some examples of AONB areas south of the A27 but which are still included are Toad's Hole Valley and Benfield Valley. • **Tide Mills, Newhaven** The Tide Mills area of Newhaven does not meet the criteria due to influence of industrial development on its eastern side. • **Lower Hoddern Farm, Peacehaven** Part of this area meets the statutory criteria for designation. However, land closest to the urban edge does not. There is no clearly definable boundary on the ground that could be adopted as the draft national park boundary and would include all land that meets the criteria. The draft boundary therefore follows the existing AONB boundary. • **Review of settlements** The settlements of Rottingdean, Ovingdean and Woodingdean posses architectural, historic and cultural associations with the Downs. However, they also contain significant areas of more recent development, not in keeping with national park quality and character, and which detract from their overall quality. They are all therefore excluded. • **Exclusion of coastal SSSI** Along the coast between Brighton Marina and Newhaven is an SSSI designation reflecting the importance of the area's geology and nature conservation. This area is regarded as not suitable for inclusion, because: a) the majority of the landscape immediately to the north of it was either built up or did not meet the quality criteria for inclusion; b) there were a number of incongruous features along the coastline, in contrast to those areas of coast between Seaford and Beachy Head which are unspoilt; and c) inclusion would create a number of 'holes' around settlements such as Saltdean. • **East Brighton** The draft boundary includes the Mount Pleasant area, an extension from the AONB. However, Sheep Cote Valley, Whitehawk Hill, Whitehawk Camp and Race Hill are excluded because they did not meet the criteria primarily due to their more urban character. • **Village Way, north Brighton** In the Brighton Borough Local Plan (Adopted 1995) there is an allocation for employment development on the above site. This allocation has been carried through to the Brighton and Hove First Deposit Draft Local Plan 2000 (following the creation of the Brighton and Hove Unitary Authority). In this latter plan, the site was allocated for employment and recreation (policy SR25 stadium proposals). The draft boundary excludes the allocated employment site illustrated in the adopted plan; however, there is no physical boundary marking the eastern side of this allocation site and on this basis the whole of the area between the University and B2123 was excluded. • **Exclusion of Brighton and Sussex Universities** The universities are excluded as they do not meet the statutory criteria, especially opportunities for access and recreation. Furthermore, the universities are not widely visible from the surrounding landscape and their buildings are not of architectural merit, nor do they act as significant landmarks.

SECTION Q: Highdown Hill (Maps 36, 37)

Boundary section	Natural beauty	Recreation	Key considerations
The draft boundary follows the northern edge of the A27 and then cuts south to join Titnore Road (A2700).	**Landscape** • Forms part of the Eastern Open Chalk landscape character area. • Includes an area of elevated chalk to the south of the A27 which is strongly associated with the wider Downs landscape, both physically and visually. • Memorable views along the south coast and northwards to the wider sweeping Downs.	• Includes Highdown Hill, which is owned and managed by the National Trust and open to the public. • Includes Highdown Gardens, which are included on the Register of Historic Parks and Gardens. • Includes an area of attractive chalk downland close to centres of population.	**West Durrington Allocation** A significant area of housing and business development is allocated in the Worthing Adopted Local Plan. This area has therefore been excluded. However, in excluding this area, other designations, which lie adjacent to the allocation, including the Titnore Wood SNCI and Castle Goring Conservation Area, were carefully assessed for inclusion.
It then continues west along the A259 before heading northwards along hedgerows south of Highdown Hill to the bridleway at Ecclesden Manor.		• Highdown Hill National Trust land offers parking, visitor facilities, and interpretation of the area's history, archaeology and nature conservation. • Good access to the area from the surrounding urban areas and wider Downs.	It was concluded that on balance these areas should be excluded on the basis that once development takes place they are likely to relate more to the urban edge than to the wider South Downs landscape. Titnore Lane forms a more suitable definable boundary.
It continues north following field boundaries to the A27, heading west.	**Nature conservation** • Includes the majority of Titnore Wood SNCI, which is important for its ancient woodland. • Includes the SNCI designation associated with The Miller's Tomb on Highdown Hill. • Includes Highdown Hill SNCI chalk grassland.		**Angmering By-pass** The Worthing Adopted Local Plan illustrates an area which is a 'protected line for a new road'. No exact alignment of the road is known at this stage. The area is, however, excluded on the basis that it does not meet the designation criteria as it is not of sufficient landscape quality.
	Cultural heritage • Includes Highdown Hill Iron Age hill fort and Anglo Saxon burial ground, which is a Scheduled Ancient Monument. • Includes Highdown Gardens Conservation Area.		

SECTION R: Arundel (Maps 37, 38, 39)

Boundary section	Natural beauty	Recreation	Key considerations
The draft boundary follows the A27 from Highdown Hill west to Crossbush. Here it crosses the A284 and continues along a narrow track to Broomhurst Farm. It then follows a line across the Arun Valley flood meadows following drainage ditches, the parish boundary and hedgerows, across Ford Road and along a hedgerow to the south of Priory Farm. It continues along the southern edge of woodland before crossing Old Scotland Lane taking the B2132 north to the A27.	**Landscape** • Forms part of the Eastern Open chalk landscape and the flood meadows of the River Arun. • Contrast between the flat and open flood meadows and the sculptured landform of the chalk, which is pronounced at Arundel. • Memorable views to the chalk and the town of Arundel, with its significant landmark buildings of a castle and cathedral. • Bold and intact landscape with a strong sense of place, where town and rural countryside strongly inter-relate. **Nature conservation** • Includes SNCI and SSSIs associated with Arundel Castle. • Includes SNCI woodland south of the A27 (Binstead Wood, Paine's Wood and Little Danes Wood). • Includes SNCI water meadows to the east and Wildfowl and Wetlands Trust site north of the town. **Cultural heritage** • Includes Arundel castle and grounds, a Scheduled Ancient Monument. • Includes St Nicolas Church (built in 1380), which is unique in being both Roman Catholic and Protestant. • Includes the remains of two Augustinian Priories at Priory Farm and Arundel station. • Includes the medieval market town and inland port of Arundel, which grew up outside the castle walls. The urban form still reflects its early role as a market and port, but most of the buildings are Georgian.	• Includes the popular River Arun footpath from Arundel onto the water meadows south of the town. • Includes the Monarch's Way long distance footpath, which passes through Arundel. • Includes the town of Arundel, which has a number of outstanding visitor attractions including the castle. This contains important treasures such as a prayer book used by Mary Queen of Scots and portraits by artists such as Gainsborough. The town also contains St Nicolas Church, as well as a significant number of buildings of architectural interest, cobbled streets, a crafts museum and heritage centre. • The town of Arundel offers a variety of tourist facilities including shops, restaurants, pubs and accommodation, as well as a tourist information centre. It is already a popular tourist destination. • Includes Arundel railway station. • Provides opportunities for walking, fishing, orienteering courses, birdwatching and site seeing. • The Wildfowl Wetlands Trust site is excellent for birdwatching and includes a viewing gallery and visitor centre. • Arundel Festival is an annual event on August Bank Holiday, and includes arts, drama, and concerts. • Arundel is closely associated with the Downs economically, hosting a local market selling local produce twice a month • Boating and cruises take place on the river from the town quay	**• Inclusion of Arundel Town** Arundel is situated on raised land above the River Arun meadows at the head of Arun Gap where the river breaks through the chalk. It is a unique, small hilltop town comprising a collection of outstanding monuments and buildings in a setting that is second to none. An assessment of the town revealed that most of it is designated a Conservation Area and Scheduled Ancient Monument, and that the historic core of the market town remains remarkably intact, with a strong sense of place. There is some more recent development south of the town but this does not adversely affect the sense of arrival at an historic place, and is not especially visible from the surrounding landscape. The castle (originally built in the 11th century and restored in the 18th and 19th centuries) dominates the skyline, particularly when viewed from the Arun floodplain. The meadows of the Arun both to the north and south of the town are a part of its landscape setting and contribute to its sense of place. The water meadows are considered to meet the statutory designation criteria, although further south the association of the water meadows to the town and to the chalk becomes less strong. For this reason, the draft boundary for the national park does not extend as far south as Tortington. **• Arundel by-pass (existing and proposed)** Within the Arun Adopted Local Plan (1993) a 'protected line for a new road' (Arundel by-pass) is illustrated. As the need for a by-pass, and its route are currently being reviewed by the Highway Agency, the area has not been excluded. The existing by-pass to the town was also assessed to determine whether it could form a suitable boundary. Adverse visual impact of the by-pass on the surrounding meadows, particularly when viewed from the south, is limited. The meadows to the south of the by-pass are regarded as meeting the criteria. They perform an important role in providing an attractive setting to the town, borrowing sense of place from the distinctive town skyline and contributing and sense of place to the town itself. Town and country are closely associated in this particular situation. On this basis, and on the basis that the meadows to the south of the town meet the statutory criteria, the existing by-pass was not regarded as a suitable boundary. The draft boundary is therefore drawn further south.

SECTION S: Slindon to West Lavant (Maps 39, 40)

Boundary section	Natural beauty	Recreation	Key considerations
From the A27 the draft boundary follows the A2132 north, before heading west along the A29, cutting through the Slindon Estate.	**Landscape** • Forms part of the transition between the Central Wooded Chalk Uplands and the Coastal Plain landscape character areas. • The Wooded Chalk Uplands is a distinctive and memorable landscape of steep-sided rolling hills, areas of significant beech and coniferous woodland and numerous estates and parklands.	• Includes the attractive Slindon Estate, which is owned by the National Trust and open to the public. • Includes Goodwood Park, a popular tourist destination. • Slindon village offers a variety of visitor facilities, including camping accommodation, pubs and local shops. • High density of public rights of way through area.	**National Trust Estate at Slindon** The draft boundary splits the existing Slindon estate in two. While the woodland south of the A29 is very similar in character and quality to that within the rest of the estate to the north, its connection to the wider estate is by only one public right of way. The A29 is preferred as the draft national park boundary. Between Duke's Road and Thicket Lane the draft national park boundary follows lanes to the south and includes land around Hungerdown, which is regarded as meeting designation criteria.
It then continues along Duke's Road, south down Britten's Lane, west along Black Mill Lane, north up Norton Lane and then west along Thicket Lane. It then runs south along the A285 until it meets the Lavant Straight where it goes westwards towards Lavant, excluding the villages of Halnaker and Boxgrove.	**Nature conservation** • Includes Slindon Wood SNCI. **Cultural heritage** • Includes Goodwood Park, registered as an Historic Park and Garden. • Includes the historic village of Slindon, with its typical brick and flint houses and associated estate, old park boundary and beech woodlands. • Includes Selhurst Park and the ancient settlement at Halnaker Hill (a Scheduled Ancient Monument). • Includes a section of the Devil's Ditch earthwork, a Scheduled Ancient Monument. • Includes the Trundle Iron Age Hillfort, a Scheduled Ancient Monument.		**Land south of the AONB** Some land to the south of the existing AONB is influenced by existing or former mineral extraction sites e.g. at Boxgrove Common, or by the gradual transition to a less distinctive and more ordinary countryside, as found to the east and west of Boxgrove. In these situations, the land is not regarded as meeting the statutory criteria and is excluded. The AONB boundary splits the settlement of Halnaker. A review of this village and that of Boxgrove concluded that neither met the criteria for inclusion, nor did the landscape surrounding these settlements merit inclusion. Although Boxgrove does have an attractive priory ruin owned by English Heritage, the settlement itself has undergone change and fragmentation of its historic core. The presence of the priory was not regarded as a strong enough reason to include the whole settlement.

SECTION T: West Chichester (Maps 40, 41, 42)

Boundary section	Natural beauty	Recreation	Key considerations
Through this section the draft boundary crosses the River Lavant south of East Lavant and then the southern and western built edge of Mid Lavant. It continues along the Devil's Ditch earthworks and then the field boundaries to Lye Lane. From here, it heads south across the B2178 and then along the northern edge of woodland and lanes to the south and west of West Ashling. It then follows Watery Lane north and the southern built edge of Funtington, before rejoining the B2178 (Hares Lane). At Racton Park Farm it heads south along field boundaries and Foxbury Lane to Westbourne, where it follows the built edge to include the River Ems valley. Finally it heads north along field edges to join Emsworth Common Road.	**Landscape** • Part of the transitional landscape between the Central Wooded Chalk Uplands and Coastal Plain landscape character areas. • Includes the valley landscape associated with the River Ems as far as Westbourne. This is a typical chalk valley landscape, of high landscape quality and unspoilt character. • Includes the attractive woodlands and parkland landscapes found around East and West Ashling, which have a strong visual and character association with the Downs. **Nature conservation** • Includes Stanstead Forest SNCI. • Includes SNCI site at Westbourne. **Cultural heritage** • Includes the historic villages of Funtington, West Ashling, East Ashling, East and Mid Lavant, all of which have Conservation Areas. • Includes Ashling Park, currently being restored. • Includes significant sections of the Devil's Ditch earthworks, a Scheduled Ancient Monument.	• An attractive and quiet rural landscape, within easy reach of Chichester and Emsworth. • There is a good network of public rights of way through this rural landscape, and many wooded areas. • This landscape forms a high-quality fringe to the steep slopes of the Downs, providing varied recreational experiences. Includes the very attractive and typical chalk downs villages, which offer local facilities and contain attractive local vernacular buildings.	• **Splitting of settlements** The existing AONB currently splits the settlements of Funtington, West Stoke and Mid and East Lavant. In all cases the settlements are regarded as worthy of inclusion and the boundary was therefore taken around the edge of the built-up area. • **Exclusion of mineral workings** The following mineral workings were identified within this area from reference to West Sussex Minerals and Waste Local Plan (March 2000): Lavant re-opening (proposed), Slades Field (proposed), Woodmancote (existing and former workings with planning permission for extension). These sites were excluded on the basis that the land did not meet the criteria.

SECTION U: Southleigh Forest to Soberton (Maps 42, 43, 44, 45, 46)

Boundary section	Natural beauty	Recreation	Key considerations
The draft boundary follows Emsworth Common Road west and then goes north to include Shuffles Plantation, following the AONB boundary. It then joins Woodberry Lane north to Rowland's Castle. It then runs along the built edge and north up Finchdean Road, joining the railway line. It then follows the built edge south, until joining Wellsworth Lane. It then joins Treadwheel Lane to the junction with Woodhouse Lane where it goes north to Blendworth Farm, including the village of Blendworth and associated parkland. From here, the boundary follows the A3 and northern edge of Clanfield before continuing south along Downhouse Road and Catherington Hill. It uses the back of settlements and field boundaries at Catherington Down. It then continues along Day Lane, Broadway Lane, Hambledon Road, Habens Lane and Roys Lane until it reaches Soberton.	**Landscape** • Forms part of the Central Wooded Chalk Uplands and more open Western Chalk Uplands. • Includes areas of strong and typical chalk landform, open rolling downs under arable cultivation or, on the steeper slopes, chalk grassland. • Includes large areas of high-quality woodland associated with Queen Elizabeth II Forest and Stanstead Forest, both with public access. **Nature conservation** • Includes Cherry Row and Blendworth woodlands, designated SINCs. • Includes Catherington Down, designated an SSSI and Local Nature Reserve for its significant area of chalk grassland and associated flora and fauna. **Cultural heritage** • Includes medieval/Norman strip lynchets at Catherington Down. • Includes Blendworth Lodge and Park, which dates from about 1810 and is currently being restored. • Includes Idsworth House and Park. • Includes Stanstead Forest/Park, registered as an Historic Park and Garden.	• Includes significant areas of parkland and woodland with public access that provide a contrast to the more open arable landscape, providing a rich variety of recreation. • Includes the extensive area of chalk grassland at Catherington Down, which has open access. • Includes Sussex Border Path, Staunton Way and Monarch's Way, all of which pass through this landscape. • Includes the Western Downs cycle route east of Rowland's Castle. • Includes the Queen Elizabeth II Country Park to the north of Clanfield.	• **Rowland's Castle** At Rowland's Castle, the draft national park boundary includes the whole of the Stanstead Park registered Historic Park and Garden and the Sussex Border Path to the north of Rowland's Castle, previously excluded from the AONB. Rowland's Castle was not regarded as meeting the designation criteria because of more recent development and because it is not surrounded by high-quality landscape. Exclusion from the proposed national park does not prevent it from acting as an important gateway into the national park. • **Areas of distinctive landscape** Areas of distinctive and high-quality chalk landscape which provide immediate and easy access from local settlements were included, e.g. North of Blendworth, north of Clanfield and Catherington Down. • **Southleigh Forest** At Southleigh Forest the draft boundary follows the AONB boundary, splitting the forest. Further south, the quality of the forest is adversely influenced by an existing landfill site and did not merit inclusion.

65

SECTION V: Meon Valley & Forest of Bere (Map 46)

Boundary section	Natural beauty	Recreation	Key considerations
The draft boundary follows the road south from Soberton on the eastern valley side of the Meon, excluding settlement associated with Soberton Heath. It then picks up and follows the Forestry Authority West Walk boundary, excluding Hundred Acres, and continues along the B2177 to Wickham. It then takes the disused railway north to join Frith Lane and Newmans Hill, before heading west along Holywell Road and field boundaries to Swanmore.	**Landscape** • Comprises the Meon Valley as it enters the Forest of Bere landscape character area and associated adjacent woodland. • The Meon Valley character continues from the chalk to form a highly attractive, and relatively enclosed, tranquil environment. **Nature conservation** • Includes Site of Importance for Nature Conservation (SINC) woodland designations associated with the Meon Valley and West Walk, e.g. Close Wood and Bishops Wood. **Cultural heritage** • Includes Rookesbury Park.	• Includes the Meon Cycle Route along the Meon Valley Railway Line, from Soberton as far as the town of Wickham. • Includes extensive areas of woodland managed by Forest Enterprise for recreation and thus including facilities such as parking, picnic areas, toilets and cycle hire. • Wickham town, although excluded, forms an excellent gateway to the proposed national park, offering a wide range of visitor facilities including shops, pubs and accommodation. • Wayfarers Walk long distance path passes through this landscape.	• **Frith Lane, Wickham** This is an active sand extraction site and landfill. Phase 1 of its restoration plan is now completed but it is likely that the site will continue to be worked for some time. It is therefore excluded.

SECTION W: Swanmore to Colden Common (Maps 46, 47, 48, 49, 1)

Boundary section	Natural beauty	Recreation	Key considerations

The draft boundary follows the built edge of Swanmore and Bishop's Waltham. It then continues west along the northern side of the B2177 to Marwell Manor.
 	
From here it adopts the edge of woodland and field boundaries to Park Farm. It continues along a track to cross the B3354 and join the A335.
	Landscape
	• Comprises the Western Chalk Uplands landscape character area, where it undergoes transition from the chalk to the coast. It is therefore a complex landscape.
	• Includes land which is typical of the Western Chalk Uplands - broad, sweeping chalk hills - bringing the national park to the edge of Bishop's Waltham and Swanmore.
	• Comprises an attractive mixture of arable land and typical beech woodland blocks.
	• Steeper slopes of remnant chalk grassland.
	• Deeply rural lanes, often tree lined, with high hedges and memorable views to the rolling chalk hills.
	
	Nature conservation
	• Includes Park Copse at Colden Common, designated as a SINC.
	• Includes Galley Down Woodland SSSI.
	
	Cultural heritage
	• Includes the historical villages of Upham and Owslebury, and notable vernacular barns at Hensting Farm.
	• Includes Marwell House registered as an Historic Park and Garden.
		• Includes Northbrook Vineyard north of Bishop's Waltham.
		• Includes the Monarch's Way and King's Way long distance footpath.
		• Includes significant area of attractive downland with a high density of public rights of way. Bishop's Waltham, although excluded from the national park, provides a range of visitor facilities and attractions, including Bishop's Waltham Palace and The Moors Nature Reserve.
			• **Bishop's Waltham**
			Bishop's Waltham has an attractive historic core and a strong sense of history.
			   However, it is not regarded as meeting the criteria because of more recent development and because it is not surrounded by high-quality landscape. It is therefore excluded.
			
			• **Marwell Manor Farm and Historical Parkland**
			Although Marwell Manor (a medieval moated site) is of historical interest, with strong connections to other Downs villages/estates, and is a Scheduled Ancient Monument, much of its historical integrity has been eroded. It therefore does not meet the designation criteria and is excluded.
			
			• **Colden Common**
			Colden Common is not regarded as meeting the criteria and some areas of landscape to the west of the settlement are regarded as more urban in character, e.g. the fish ponds.
			   The draft boundary is therefore kept back from the urban edge, but includes Park Copse, an attractive woodland designated a SINC.

9 The next steps

This consultation is about the proposed administrative arrangements for a South Downs National Park Authority and the draft boundary for a South Downs National Park. We would like your views on the key issues and the draft boundary. See page 8 for details of how and where to send your comments.

After considering your responses, the Countryside Agency will prepare a proposed boundary for a South Downs National Park and its recommendations for government on the arrangements for a South Downs National Park

Authority. It will consult formally with local authorities (including parish and town councils) on the proposed boundary, as required by legislation, and will also invite their comments on its recommendations to government on the national park authority in spring 2002. The Countryside Agency will then make the National Park Designation Order and submit it to government, along with its recommendations, in summer 2002. The Order will be published and the recommendations will be made public.

The Secretary of State responsible is likely to announce in late 2002 whether he/she will confirm the Designation Order or whether a public inquiry is to be held. The Secretary of State may also indicate then his/her intention to establish a national park authority, and any special arrangements to be put in place. If a public inquiry is not required, a final decision on the Order by the Secretary of State might be taken in spring 2003. If a public inquiry takes place during 2003, it might well be 2004 before final decisions are taken.

10 Key steps for establishing a South Downs National Park

(under the National Parks and Access to the Countryside Act 1949, as amended by the Countryside Act 1968 and the Environment Act 1995)

Creating the National Park

Step 1: Spring 2000
Countryside Agency (CA) decides to begin process of designating the South Downs as a National Park.

Step 2: Spring 2001
CA identifies 'area of search' and begins work on defining a draft boundary.

Step 3: Autumn 2001
CA undertakes 3 month public consultation on draft boundary.

Step 4: Spring 2002
CA considers responses to public consultation and decides and decides on changes to the draft boundary.

Step 5: Spring 2002
CA undertakes 'formal' consultation on proposed boundary with local authorities affected, as legislation requires.

Step 6: Summer 2002
CA considers local authority responses to boundary consultation, decides what area should be designated as a National Park and makes the Designation Order. The Order includes a map and description of the area to be designated.

Step 7: Summer 2002
CA advertises in local papers and the London Gazette that the National Park Designation Order has been made, where it can be inspected and for how long.

Step 8: Autumn 2002
At least 28 days are allowed for representations or objections to the Order to be made to the Minister.

Step 9: Winter 2002
Minister considers any objections or representations to the Designation Order and decides whether to call a public inquiry. (He/she will do so if there are objections by any local authority).

Step 10: 2003/04?
Minister decides whether to confirm Designation Order, possibly after a public inquiry. If Designation Order is confirmed, South Downs National Park is created.

Administrative arrangements for a South Downs National Park Authority (NPA)

Step 1: Spring 2000
CA discuss issues and possible solutions with key bodies and individuals at meetings, technical working groups and seminars, and seeks specialist advice where necessary.

Step 2: Spring 2001
CA Publishes 'Mission Statement' for a South Downs National Park Authority and develops options for administrative arrangements.

Step 3: Autumn 2001
CA undertakes 3 month public consultation on options for administrative arrangements.

Step 4: Spring 2002
CA considers responses to consultation on options for administrative arrangements and decides on proposed advise to government.

Step 5: Spring 2002
CA seeks comments from local authorities on proposed advise to government.

Step 6: Summer 2002
CA recommends to Government administrative arrangements for a South Downs NPA.

Step 7: Winter 2002
Government considers CA advice.

Step 8: 2003/2004?
If Designation Order is confirmed, Government makes establishing Order creating a South Downs NPA and may issue guidance to the NPA (eg. a circular).

Appendices

Appendix A.
National park legislation

National parks are the most beautiful expanses of country in England and Wales where people can enjoy a wide range of open-air recreation. The term National Park is a statutory title that recognises an area's national importance and provides the highest degree of protection for its landscape.

Proposals for national parks in England and Wales were first made in 1945 when the Government published a White Paper[22], written by John Dower. He gave shape to the concept of national parks that remains valid today. He defined a national park as: "an extensive area of beautiful and relatively wild country in which, for the nation's benefit and by appropriate national decision and action:

- the characteristic landscape beauty is strictly preserved;
- access and facilities for open-air enjoyment are amply provided;
- wildlife and buildings and places of architectural and historic interest are suitably protected;
- while established farming use is effectively maintained".

Dower's recommendations were followed in 1947 by the Report of the National Parks Committee (commonly known as the Hobhouse Report, after its chairman, Sir Arthur Hobhouse)[23]. It recommended which areas should be selected as national parks and proposed an administrative system for them.

Legislation, purposes and criteria

National parks are designated under the National Parks and Access to the Countryside Act 1949. This Act implemented the findings of the Hobhouse Report, making provision for national parks and establishing a National Parks Commission.

The 1949 Act was amended by the Countryside Act 1968 and the Environment Act 1995. The table overleaf gives the national park purposes and criteria from Section 5 of the Act and the definition of natural beauty from Section 114 as amended by the 1968 and 1995 Acts. This is the statutory basis for designating national parks.

In 2000, in response to a request from the Government, the Countryside Agency reviewed the application of the criteria and decided that the key questions to be considered in designating new national parks (other than natural beauty) were[24]:

Is it an extensive tract of country providing or capable of providing sufficient opportunities for open air recreation?

Is it especially desirable to provide for the leadership of a national park authority, with the powers and duties laid down in the Environment Act 1995?

22 Dower, J, National parks in England and Wales, HMSO, 1945.
23 The National Parks Committee, Report of the National Parks Committee (The Hobhouse Committee), Cmd 7121, HMSO, 1947.
24 Letter from Ewen Cameron, Chairman of the Countryside Agency, to the Rt Hon Michael Meacher, Minister for the Environment, 16 March 2000.

Extracts from the 1949 Act as amended by the 1968 and 1995 Acts

Statutory national park purposes and criteria (Section 5)

1   The provision of this Part of this Act shall have effect for the purpose -

a   of conserving and enhancing the natural beauty, wildlife and cultural heritage of the areas specified in the next and following subsection; and

b   of promoting opportunities for the understanding and enjoyment of the special qualities of those areas by the public.

2   The said areas are those extensive tracts of country in England and Wales as to which it appears to the Commission that by reason of -

a   their natural beauty, and

b   the opportunities they afford for open-air recreation, having regard both to their character and to their position in relation to centres of population, it is especially desirable that the necessary measures shall be taken for the purposes mentioned in the last foregoing section.

The said areas, as for the time being designated by order made by the Commission and submitted to and confirmed by the Secretary of State, shall be known as, and are hereinafter referred to as, National Parks.

Definition of natural beauty (Section 114)

References in this Act to the preservation, or conservation of the natural beauty of an area shall be construed as including references to the preservation or, as the case may be, the conservation, of its flora, fauna and geological and physiographical features.

Appendix B.
History of protection and designation in the South Downs

1923   Society of Sussex Downsmen formed

1934   South Downs Preservation Bill proposed by local councils but not made law

1935   The Standing Committee on National Parks established by the CPRE (England and Wales)

1945   Government report on national parks in England & Wales written by John Dower

1947   Report of the National Park Committee chaired by Sir Arthur Hobhouse to look at setting up national parks in England identifies 12 potential national parks, including the South Downs

1949   National Parks & Access to the Countryside Act passed: first national park (Peak District) designated in 1951

1957   National Parks Commission considers a South Downs National Park, but rejects it because:
"The recreational value of the South Downs as a potential national park has been considerably reduced by extensive cultivation"

1962   East Hampshire designated as an AONB

1966   Sussex Downs designated as an AONB

1972   South Downs Way National Trail designated (Sussex part)

1986   Sussex Downs Forum established by the Sussex local authorities.

1987   Agriculture Act passed - South Downs Environmentally Sensitive Area created in order to conserve and enhance the landscape, historic and wildlife value of the Downs

1990   South Downs Campaign formed to press for a South Downs National Park

1991   South Downs Way extended to Hampshire

1991   East Hampshire Joint Advisory Committee (JAC) established

1991   Sussex Downs Conservation Board (SDCB) established as a six-year national experiment in managing AONBs under special pressure

1994   East Hampshire JAC appoints a project officer to implement the management plan

1998   Countryside Commission reviews the management of the South Downs and concludes that it will be best served by a new statutory Conservation Board with greater powers and independence which would require new legislation.

1998   SDCB experiment extended for a further three-year period

Feb 2000   Countryside Agency reviews its policy for the interpretation of national park criteria, taking account of modern recreational needs

April 2000   Countryside Agency decides to begin the designation of a South Downs National Park

Circular 12/96
(Department of the Environment)

Circular from the

Department of the Environment
2 Marsham Street, London SW1P 3EB

11 September 1996

Environment Act 1995, Part III
National Parks

Introduction

1.   I am directed by the Secretary of State for the Environment to draw your attention to the action that is necessary to implement the provisions of Part III of the Environment Act 1995 (hereinafter referred to as the 1995 Act) as they relate to the administration of National Parks in England.

2.   The 1995 Act enables authorities (hereinafter referred to as the "National Park Authorities") to be established to administer the National Parks in England and Wales. It also introduces revised purposes for the National Parks and imposes new duties on both the National Park Authorities, when established, and on other public bodies operating in the Parks.

Background

3.   "Fit for the Future", the 1991 report of the National Parks Review Panel, defined the essence of the concept of National Parks as lying in "the striking quality and remoteness of much of their scenery, the harmony between man and nature it displays, and the opportunities it offers for suitable forms of recreation". This definition was endorsed by the Government in its 1992 policy statement in response to "Fit for the Future".

4.   National Parks were designated under the provisions of the National Parks and Access to the Countryside Act 1949 for the twin purposes of preserving and enhancing their natural beauty, and of promoting their enjoyment by the public, as set out in section 5 of the 1949 Act. The twin purposes of conservation and public enjoyment have largely stood the test of time, but the Government's 1992 statement agreed that some revision was needed.

5.   The addition of a third purpose, to promote the economic and social development of the communities within the Parks, was also considered by

1

the Review Panel, but rejected. The Government agrees. It is not for the National Park Authorities to assume the role of those development agencies which already exist for these purposes. However, it is essential that the National Park Authorities take full account of the economic and social needs of local communities in fulfilling the purposes of the Parks and this can only be achieved by working in close co-operation with local authorities, landowners and land managers and those other agencies and persons with interests in the Parks. (See paragraphs 25-29).

6. The Government regards National Park designation as conferring the highest status of protection as far as landscape and scenic beauty are concerned. The national significance which led to the designation of the Parks continues to be reflected in both the funding and membership arrangements for National Park Authorities. Three quarters of the National Park Authorities' approved expenditure comes direct from central Government; the remaining quarter via local authorities. Similarly, one half plus one of each National Park Authority's members are appointed by the constituent local authorities. The remainder are appointed by the Secretary of State, of whom one half minus one are nominated by parish councils in the Park (see paragraphs 30-41).

National Park purposes

7. The revised National Park purposes, as set out in section 61 of the 1995 Act, are:

(i) to conserve and enhance the natural beauty, wildlife and cultural heritage of the National Parks; and

(ii) to promote opportunities for the understanding and enjoyment of the special qualities [of the Parks] by the public.

These revised purposes are an explicit expression of the wider-ranging view which the National Park Authorities should take, and which the National Park Boards and Committees have previously reflected in the integrated planning and management of their Parks. The revision of the purposes takes into account the changing nature of the Parks and the pressures upon them whilst ensuring that the conservation values which the Parks represent continue to be fully reflected. The new purposes in themselves are not, however, expected to result in a major change to the way the National Parks in England operate.

Conserving and enhancing the natural beauty, wildlife and cultural heritage of the National Parks

8. The National Parks are areas of exceptional natural beauty. They contain important wildlife species and habitats, many of which have been designated as being of national and international interest. But the Parks are also living and working landscapes and over the centuries their natural beauty has been moulded by the influence of human activity. Their character is reflected in local traditions which have influenced farming and other land management practices. It is also reflected in the local building materials and vernacular style, monuments and landscape, often of archaeological or historical significance, and in the words, music, customs, crafts and art

2

which mark the individual characteristics of each Park. The Parks represent an important contribution to the cultural heritage of the nation.

9. As they develop and implement policies for the management and planning of their areas, the National Park Authorities will need to determine how best to reflect the qualities of natural beauty, wildlife and cultural heritage which they are to conserve and enhance. The National Park Authorities will be particularly expected to respond to and seek to conserve the individual character of the area for which they are responsible. In meeting the revised conservation purpose, the National Park Authorities will be expected to work closely with appropriate bodies including local authorities, the Countryside Commission, MAFF, English Nature and English Heritage.

Promoting opportunities for the understanding and enjoyment of the special qualities of National Parks by the public

10. The attraction of the special qualities of the National Parks has long been recognised and is demonstrated by the numbers of visitors who seek the enjoyment of these beautiful areas and the opportunities they afford for open-air recreation. In setting out the criteria for designating National Parks, the 1949 Act recognised the opportunities the Parks provide in this respect.

11. The Government believes that individual National Park Authorities are best placed to identify the nature of the special qualities of their Parks. These qualities will be determined within the context of each Park's natural beauty, wildlife and cultural heritage and the national purpose of the Parks to conserve and enhance them. Particular emphasis should be placed on identifying those qualities associated with their wide open spaces, and the wildness and tranquillity which are to be found within them.

12. The National Park Authorities should consider how best to promote the understanding of the special qualities of their areas by the public. They should produce strategies to promote this understanding, which may involve the use of the Parks' own specialist staff and services, such as rangers, wardens and education officers, publicity and educational material, and information and study centres. It will also involve working in partnership with other organisations, including local education authorities; schools, colleges and their teachers; training organisations; tourist boards and tourist operators; relevant Government bodies; local and national amenity societies; and writers, artists, publishers and broadcasters.

13. The National Park Authorities should continue to promote the widest range of opportunities for recreation to reflect the variety of ways in which the Parks can be enjoyed. But the conservation values which the Parks represent and which have led to so many appreciating their special qualities must be fully respected. The National Park Authorities will need to take into account the Parks' limited environmental capacity. It will not be appropriate for all forms of recreation to take place in every part of the Parks and the Government accepts that some recreational activities could cause unacceptable damage or disturbance to their natural beauty, wildlife or cultural heritage. The intrusive nature of some recreational activities or the damage they cause to conservation interests may unacceptably affect other people's understanding and enjoyment of the Parks.

3

14. Nevertheless the Government does not accept that particular activities should be excluded from throughout the Parks as a matter of principle. They contain a variety of landscapes, capable of accepting and absorbing many different types of leisure activity. In most instances, it should be possible to reconcile any conflict which may arise by co-operation between relevant interests and the National Park Authorities, and through careful planning and positive management strategies.

15. National Park Authorities will be expected to take a positive role in implementing the revised second National Park purpose and should play a proper role in promoting those opportunities. They should work closely with the Countryside Commission, the Sports Council, the Tourist Boards, the Forestry Commission and English Nature to achieve this. The Government believes that the promotion of the second purpose is not incompatible with the equally legitimate demand that many parts of the Parks should continue to be quietly enjoyed by many people for much of the time. It further believes that co-operation is the best means of ensuring sensitive use of the National Parks. Nor must the second National Park purpose be interpreted so as to imply undue interference with the everyday lives of those who live and work in the Parks. It is of particular importance that those experiences which are unique to the National Parks should be protected and fostered.

Sandford principle

16. "Fit for the Future" strongly endorsed the recommendation of the Sandford Committee in 1974 that the statutory purposes of the Parks should make it clear that their enjoyment by the public "shall be in a manner and by such means as will leave their natural beauty unimpaired for the enjoyment of this and future generations". The Sandford Committee concluded that most conflicts could be resolved by good management, but stated that "where it is not possible to prevent excessive or unsuitable use by such means, so that conflict between the two purposes becomes acute, the first one must prevail in order that the beauty and ecological qualities of the national parks may be maintained." The intention behind this principle, which has been Government policy for over 20 years and is known as the "Sandford principle", is now enshrined in section 62 of the 1995 Act. The Government believes that it continues to stand the Parks in good stead.

17. The National Park Authorities and other public bodies, as they exercise their functions in the National Parks, should make every effort to reconcile any conflicts which may arise between the two National Park purposes. The National Park Authorities in particular will be expected to encourage mediation, negotiation and co-operation, but there may be instances where reconciliation proves impossible. In those cases, the conservation purpose should take precedence. For example, in cases where excessive visitor pressure, or a particular type of activity, is likely to destroy or degrade, some management of access may be necessary, otherwise there may be nothing left to conserve or to enjoy. As the National Park Authorities are ultimately responsible for ensuring that National Park purposes are met, it is for them to determine when conflict arises between the two Park purposes, although it is important for them to keep in mind and fully consider the views of all in the Parks.

4

Sustainable Development

18. The Government is committed to the principle of sustainable development as the cornerstone of policies to reconcile the needs of economic development with those of environmental protection. The Government believes that the National Parks are in a strong position to influence the way in which we care for our countryside, to be models for the sustainable management of the wider countryside, and to help to further general understanding and appreciation of the means by which development and conservation can be better balanced. The National Park Authorities will have regard to the principles of sustainable development as they undertake their duties. Sustainable development is an important principle in achieving the well-being of local communities while aiding the conservation and enhancement of biodiversity.

The application of National Park purposes

19. Section 62 of the 1995 Act places a general duty on all relevant authorities, including the National Park Authorities, statutory undertakers and other public bodies, to have regard to the purposes of the Parks as set out in revised form in section 61. This ensures that they take account of Park purposes when coming to decisions or carrying out their activities relating to or affecting land within the Parks. Relevant authorities will be expected to be able to demonstrate that they have fulfilled this duty. They will wish to consider whether they could usefully make reference to it in their annual reports. It may sometimes be the case that the activities of certain authorities outside a National Park may have an impact within the Park. In such cases it will be important to ensure mutual co-operation across Park boundaries, particularly in planning and highway matters.

The economic and social well-being of Park communities

20. The Government is concerned to ensure that there is no incompatibility between conserving the National Parks and their remaining as living and working communities. The qualities for which the Parks have been designated are as much the products of man's hand as of nature. It is in the interests of the conservation of those qualities that the National Park Authorities have a duty to work with and for their local communities.

21. For these reasons, in pursuing the purposes of the Parks, section 62 of the 1995 Act places on the National Park Authorities a duty to seek to foster the economic and social well-being of their local communities. This new duty reinforces the need for the Parks to take a positive view of the well-being of their local communities, although it does not enable the National Park Authorities to make additional financial resources available and gives them no new powers, except in relation to National Park purposes. The National Park Authorities will need to consult MAFF and the Forestry Commission over the socio-economic effects of their policies.

22. The Government expects the National Park Authorities to ensure that they carry out the duty to foster the economic and social well-being of their local communities in ways which are compatible with their pursuit of National Park purposes and with the need for appropriate economic development. The natural disadvantages which may exist for those living in

5

National Parks, such as remoteness, climate and geography, are well recognised and are legitimate concerns to be taken into account by the National Park Authorities as local planning authorities. However, the economic advantages resulting from National Park designation should also be emphasised: tourism may result in significant benefits in terms of local purchasing and employment and is also supported by the National Park Authority through the operation of its visitor services; areas within the Parks qualify for funding under Objective 5b and the Rural Development Programme; and farmers and landowners receive assistance from the Park Authorities in their stewardship of land within the Parks.

23. The National Parks are part of the social, economic and environmental structure of a wider area. The Government expects the National Park Authorities to continue to co-operate with local authorities, and other agencies, such as the Rural Development Commission, whose task it is to promote the social or economic development of rural areas.

24. It is the Government's view that it is not appropriate for the National Park Authorities themselves to assume the role of promoting economic and social development in the Parks, nor to compete with those agencies which have the power to do so. Section 62 of the 1995 Act therefore prevents National Park Authorities from incurring significant expenditure in carrying out their duty towards the social and economic well-being of local communities, for example, expenditure other than appropriate administration costs resulting from liaison with local authorities and other agencies. This ensures that the financial resources of the Park Authorities remain focused on those purposes for which the Parks were created, but does not constrain any other of the Parks' powers to aid local interests through grants or management agreements. Expenditure in support of National Park purposes which has the indirect benefit of fostering the economic well-being of local communities is entirely proper provided that the expenditure is primarily incurred for those National Park purposes.

Consultation with national and local interests

25. In formulating policies for the administration and management of the Parks, the National Park Authorities must have in mind the wide range of interests which can be affected by their decisions. These interests include those of people who live and work in the Parks, those whose living is derived from the resources of the Parks, and those who visit the Parks.

26. The National Park Authorities will especially be expected to work closely with the Countryside Commission, which has statutory responsibilities in respect of National Park designation and boundary variation procedures and in advising the Secretary of State on National Park financial and membership matters. They will also be expected to draw on the knowledge and expertise associated with the statutory responsibilities of other national agencies: in particular, in the light of the revised first National Park purpose, of English Nature and English Heritage; and in the light of the second purpose, of the Sports Council, the Central Council for Physical Recreation and the regional Tourist Boards. They will also be expected to work in close co-operation with the Environment Agency established under Part I of the 1995 Act, all relevant Government organisations including

6

MAFF and the Forestry Commission, and appropriate commercial bodies and non-governmental organisations.

27. It is especially important that the National Park Authorities should work closely with their constituent local authorities in matters which impinge on their responsibilities. The Government urges both the National Park Authorities and the local authorities within whose boundaries the Parks lie to seek effective co-operation which fosters the health and vitality of local communities whilst respecting the special qualities of the Parks.

28. The Government in turn looks to all these agencies and organisations to involve the National Park Authorities fully in their activities as they affect the Parks.

29. The Government also emphasises the importance of ensuring that the views of local people are fully considered. All National Park Authorities should make formal arrangements to ensure that local people as well as local voluntary groups and recognised user interests in the Parks, have a proper and regular opportunity to make known their views. Paragraph 16 of Schedule 7 to the 1995 Act requires National Park Authorities to make particular arrangements with each parish council, whose area lies wholly or partly within the Park, to inform and consult them about the Authority's discharge of its functions. Collective consultation arrangements are generally to be encouraged.

Membership

30. The National Park Authorities (England) Order 1996 (SI 1996 No. 1243), made by the Secretary of State under Section 63 of the 1995 Act, sets out the membership distribution of each National Park Authority. Before making the order, the Secretary of State was required to consult the principal local authorities about the composition of the Authority, including the overall number of members and the number of members which each local authority should appoint. Schedule 7, paragraph 2(3) to the 1995 Act makes provision for the Secretary of State to consider the exclusion of a council from membership of the National Park Authority only at the request of that council. Mid Devon District Council has been so excluded in respect of Dartmoor.

31. Local authorities appoint one half plus one of the members of the National Park Authorities. The remainder are appointed by the Secretary of State, of whom one half minus one are parish representatives. Training is to be offered to all appointees by National Park staff, assisted as appropriate by the Countryside Commission.

Allowances and time off for members

32. Paragraph 11 of Schedule 7 to the 1995 Act states that the provisions of the Local Government Act 1972 and the Local Government and Housing Act 1989 regarding allowances for travelling, conferences and visits and basic attendance and special responsibility allowances respectively, shall apply equally to Secretary of State members of National Park Authorities as they do to local authority members. The payment of allowances is the responsibility of the National Park Authority, which will be required to make an annual statement of such payments.

7

Local authority appointments

33.   Paragraph 2 of Schedule 7 to the 1995 Act requires local authority appointees to be serving councillors of their appointing local authority, and also requires local authorities to have regard to the desirability of appointing members who represent divisions or wards situated wholly or partly within the relevant Park. They should also have relevant experience and close links to the Park. In determining the overall size of the National Park Authorities, the Secretary of State has taken into account the requirements of the 1995 Act to ensure that every relevant local authority is represented. In areas where two tier principal local authorities exist, the Secretary of State has ensured that the total number of members from each tier is equal. In conjunction with these principles, the Secretary of State believes that the membership numbers of each National Park Authority should be kept as small as possible consistent with effectiveness and an equitable distribution of local representation.

Appointments by the Secretary of State

34.   The Secretary of State will take steps to encourage all those with an interest in the Parks to make nominations for appointments to the National Park Authorities. In making his appointments, the Secretary of State will be concerned to take account of the national purposes for which the Parks have been designated. In selecting, after consultation with the Countryside Commission, persons suitable for appointment, the Secretary of State will look for a capacity to present this wider viewpoint in discussions within the Authority and for experience, preferably in a combination of fields, with direct relevance to the character of the particular National Park and to the responsibilities of the Authority. Wherever possible the Secretary of State will give preference to candidates who combine these qualities with local association to the Park to which they are appointed.

35.   Individuals will be selected for their personal qualities and experience and not as representatives of specific groups or organisations. Whilst the Secretary of State expects his members to have regard to the interests of all those concerned with a specific Park, his primary concern is that they should bring to the Authorities' deliberations the wider national viewpoint.

36.   Secretary of State members are usually appointed for a three year term, although the Secretary of State has discretion to appoint his members for shorter periods. Members are eligible for reappointment but are not usually expected to serve more than three successive terms. Nominations for appointment will be invited each year by the Department of the Environment. It should be borne in mind that very few vacancies arise each year. Nominees will remain on the list of potential candidates for appointment for three years and, unless a fresh nomination is received, will be removed at the end of that period. Those putting forward nominations should inform nominees that they have done so. The Countryside Commission has a statutory role in advising on National Park appointments and will continue to be asked to interview those candidates shortlisted by Ministers for appointment prior to a final selection by the Secretary of State.

8

37. The Secretary of State does not propose to appoint as a Secretary of State member anyone who is a serving councillor of a county or district council appointing members to the National Park Authority, or anyone employed by such a council.

Parish members

38. Paragraph 3 of Schedule 7 to the 1995 Act enables the Secretary of State to appoint parish members to the National Park Authorities. Parish members must either be members of a parish council, or the chairman of the parish meeting of a parish which does not have a separate parish council, wholly or partly situated within the National Park. Parish members of a National Park Authority are to hold office from the time of their appointment until they cease to be a member of the parish council or chairman of the parish meeting. The need to stand for re-election to the parish council or as chairman of the parish meeting will terminate their appointment to the National Park Authority, but they will be eligible for re-appointment if they are re-elected. Parish members will be appointed on the same terms and conditions, and subject to all the usual rules on conduct, as local authority and Secretary of State members of the National Park Authorities.

39. The appointment by the Secretary of State of parish members to the National Park Authorities is to ensure that local people have a greater involvement in the running of the National Parks and in the management of Park affairs. It enables a proper balance to be achieved between the wider national interest, that of local authorities and the truly local concerns of those who live and work in the Parks.

40. Parish members are appointed to represent the wider Park view and not just the interest of their own parish, and are representatives rather than delegates of the grouping of parishes nominating them. The Secretary of State looks to parishes in each National Park to maintain a local mechanism to select candidates commanding general support whom he can appoint to the Authority. He does not propose to determine the mechanism, but will look to a result which will enable the full range of areas within the Park to be represented. Where the requisite number of parish members comes forward the Secretary of State would propose generally to appoint them. In default of such agreement, the Secretary of State would select candidates from amongst the local nominees.

41. The Secretary of State does not propose to appoint as a parish member anyone who is a serving councillor of a county or district council appointing members to the National Park Authority, or anyone employed by such a council.

Finance

42. Sections 71 and 72 of the 1995 Act provide for a local and national element in the financing of the National Park Authorities. Given the present general arrangements for the financing of local authorities, the Secretary of State intends to provide 75% of the approved expenditure of the National Park Authorities, with the remaining 25% provided by those principal local authorities appointing members to the Authority. The local authorities'

9

contributions will be raised by levy issued by the National Park Authorities. Section 71(5) of the 1995 Act gives the opportunity for all the local authorities to whom a levy may be issued for a particular Park to agree before 1 December on how the levy should be apportioned for the following year.

43.   The Secretary of State is confident that the local authorities will be able to reach agreement on the proportions which each will contribute towards the total local authority contribution for National Park Authority expenditure. In the event that they cannot reach agreement, however, provision is made in the 1995 Act for the matter to be decided in accordance with regulations under section 74 of the Local Government Finance Act 1988.

Annual reports

44.   In accordance with section 230 of the Local Government Act 1972, the Secretary of State expects reports to be produced each year. The National Park Authorities will be required to produce, publish and give wide circulation to annual reports of their stewardship of the Parks. Copies should be sent to the Secretary of State and all relevant national agencies. The reports will be expected to include consideration of matters which are of interest to their constituent local authorities, parish councils and other bodies operating within the Park. These reports are important as a basis for the continuing dialogue between the National Park Authorities and their local communities.

45.   The reports should show how each National Park Authority has sought to achieve the purposes of the Park, exercised its statutory duties and implemented the policies in the National Park Management Plan. They should also include a summary of the state of the Park environment, and refer to the achievement of key Government policies. There should also be a summary financial statement, a breakdown of expenditure and details of any special achievements over the year.

Planning responsibilities of the National Park Authorities

46.   As sole local planning authority for its area under section 4A of the Town and Country Planning Act 1990, a National Park Authority is responsible for maintaining structure plan, local plan, and minerals and waste local plans coverage. If the Secretary of State so orders under section 67(2) of the 1995 Act, it will instead be responsible for preparing a unitary development plan; although the Secretary of State does not propose to make any such orders in the first instance. The National Park Authority will also exercise development control functions for its area.

47.   A National Park Authority will be the strategic land-use planning authority for its area. Unless the Secretary of State has made an order under section 67(2) of the 1995 Act, the authority will exercise that responsibility by updating or replacing the existing structure plan provisions for its area. Where it is desirable, the Authority may seek, and is encouraged, to make voluntary arrangements under section 101 of the Local Government Act 1972 with one or more neighbouring strategic planning authority to prepare a joint structure plan for their combined area. Except in the case of the Peak

10

District National Park for which a single area-wide strategic plan continues to be appropriate, the Secretary of State is looking to the National Park Authorities to work with neighbouring strategic planning authorities to maintain a joint structure plan for their combined areas. Where necessary (there is already a joint structure plan for Cumbria and the Lake District National Park), the provisions of paragraph 5 of Schedule 4 to the National Park Authorities (England) Order 1996 facilitate the desired pattern of joint structure plan working.

48. A structure plan sets out key strategic planning policies as a framework for more detailed policies in local plans, which should be aimed at guiding development in the Park. The National Park Authority will need to consult its constituent local authorities before preparing their development plan proposals. In accordance with the development plan provisions of the 1990 Act, and the provisions of the Town and Country Planning (Development Plans) Regulations 1991, it will also need to consult a number of other bodies (which are listed in Annex E to Planning Policy Guidance Note 12). These include all relevant national and local agencies and organisations and statutory undertakers, as well as Government Departments with an interest in the development plan proposals.

Major Development

49. Government planning policy towards the National Parks, as well as the Broads and the New Forest, is that major development should not take place in these areas save in exceptional circumstances. Because of the serious impact that major developments may have on their natural beauty, applications for such developments must be subject to the most rigorous examination and should be demonstrated to be in the public interest before being allowed to proceed. Consideration of such applications should therefore normally include an assessment of:

> the need for the development, in terms of national considerations, and the impact of permitting it or refusing it upon the local economy;

> the cost of and scope for developing elsewhere outside the area or meeting the need for it in some other way;

> any detrimental effect on the environment and the landscape, and the extent to which it should be moderated.

Old Minerals Permissions

50. Part V of the 1995 Act incorporates provisions requiring minerals planning authorities to prepare lists of active and dormant pre-1982 minerals sites within their areas. All sites wholly or partly within National Parks are Phase I sites where either the whole or greater part of the land is subject to relevant planning permissions granted after 30 June 1948 and before 22 February 1982. Separate guidance has been issued.

National Park Management Plans

51. Section 66 of the 1995 Act requires National Park Authorities to prepare and publish National Park Management Plans as statements of their policy for managing and carrying out their functions in relation to the

11

Parks. All National Park Authorities already have National Park Plans for their areas which perform a similar role and which are required to be reviewed every five years. Under section 66(2) of the 1995 Act a new National Park Authority will be permitted to adopt the existing plan as its Management Plan, providing that it publishes notice that it has done so. If an existing Plan is due for review within 12 months after the new National Park Authority is established, the National Park Authority may review the Plan before adopting it.

52.   At the request of the Secretary of State the Countryside Commission publishes advisory notes on the production of National Park Plans. The guidance advises on the process for producing National Park Management Plans and on their content. It emphasises the importance of working closely with appropriate interested bodies and highlights the role of the Plans as strategic documents outlining overall policies. New Plans should reflect the revised National Park purposes, the duties of the National Park Authorities and the need for statutory consultation as appropriate. They should complement the structure, local and minerals and waste local plan coverage of the Parks. They may be supported by inventories of Park resources and a range of more detailed documents dealing with specific topics, such as nature conservation or forestry or management plans for particular areas.

Roads and traffic

53.   Circular 125/77 (Department of Transport Circular 182/77) 'Roads and Traffic – National Parks' – remains extant. It advocates close working between highway authorities and the National Park Authorities, including periodic consultation on proposed road programmes and notification of all individual improvements. Highway and traffic authorities are advised to play their proper part in the implementation of National Park plans. The Circular recommends definition of a functional road hierarchy within the Parks, with appropriate traffic management measures.

54.   The policy on major developments in National Parks (paragraph 49) applies to transport developments. There is a statutory duty to consult the Countryside Commission when an environmental statement is required for a highway development within a National Park. The Government encourages the National Park Authorities to work closely with local highway and traffic authorities (including the Highways Agency) to develop appropriate schemes for traffic and transport management, in full consultation with local interests. Consultation with the Department of Transport (including the Government Office for the Region and the Highways Agency, as appropriate) is encouraged where schemes are innovative in nature.

Rights of Way Responsibilities

55.   Relevant highway authorities are asked to consider positively their relationship to the new National Park Authorities, and to work closely with them. They are urged to enter into agency agreements to delegate rights of way work to National Park Authorities where that has not already been done. Those agreements will enable National Park Authorities to continue to promote and protect footpaths and bridleways to meet the particular circumstances of their Park.

12

Defence use of National Parks

56.   Parts of some of the National Parks have a long tradition of defence use, which pre-dates the designation of the Parks by many years. While the Government accepts these existing uses will continue into the foreseeable future, it is nevertheless committed to ensuring that new, renewed or intensified use of land in the National Parks for defence purposes should be subject to formal consultation with the National Park Authorities and the Countryside Commission and to an environmental impact assessment, and should be tested against any provisions set out in planning policy guidance. It acknowledges however, that there can be conflicts between defence use and Park purposes, but believes these will be best resolved through co-operation with the National Park Authorities. The Ministry of Defence will continue to give a high priority to conservation.

57.   Defence use of National Parks makes a major contribution to the country's defence capability, and provides essential facilities which could not be easily replicated elsewhere. It can also be an important factor in contributing to the local economic and social well-being of Park communities.

Circular 4/76

58.   The guidance in this Circular replaces that contained in DOE Circular 4/76 "Report of the Review of the National Park Policies Review Committee". The Annex to that Circular, which contains the conclusions of the Government of the day on the Report of the National Parks Policies Review Committee (the Sandford Report), nevertheless remains of interest as an important contribution to the development of policy for the National Parks.

R M PRITCHARD, *Assistant Secretary*

The Chief Executive
    All County Councils, District Councils and Unitary Authorities in England whose area falls partly within the designated area of a National Park
National Park Officers in England
The Chief Executive, the Broads Authority
[DOE   CYD/4367/183]

Printed in the United Kingdom for HMSO
Dd302944   C30   9/96   2100   17434

13

Appendix D.
Technical advice on administrative arrangements

Following the decision in April 2000 to begin the process of designating a South Downs National Park, the Countryside Agency has carried out detailed work to identify administrative issues which need to be addressed in establishing a South Downs National Park Authority, and to find solutions to these.

From this the Agency will prepare its advice to government on how to set up a national park authority, and in due course to a national park authority on how to carry out its responsibilities.

The Agency's aim was to:
- take account of existing management arrangements;
- identify any gaps in existing policy and practice;
- build on what already exists - where it is working well;
- avoid unnecessary duplication and bureaucracy;
- draw on good practice from elsewhere, especially other national parks;
- take account of the particular features of the South Downs;
- take into account cross-border issues and working in partnership with organisations within and outside the proposed boundary.

The work took several forms:
- Technical advisory groups were set up to look at governance and administration; planning; and the draft boundary.
- Three topic groups were established, run on behalf of the Countryside Agency by the Sussex Downs Conservation Board and the East Hampshire Joint Advisory Committee[25]. They focussed on land management, sustainable recreation and community involvement. In addition to addressing administrative issues the groups also discussed and developed some policy ideas for a South Downs National Park Authority to consider.
- A seminar, which more than 150 delegates attended, was held to discuss the general issues and implications connected with the proposed national park and its boundary.
- Meetings took place with local and national statutory and voluntary bodies, local authorities, parish councils and farmers.
- Specialist and legal advice was sought, including advice on planning from the Department for Transport, Local Government and the Regions and other bodies with particular responsibility.

25 Members of all of these groups were chosen as individuals because of their technical expertise and/or local knowledge, and did not represent organisations.

Glossary

For the purposes of this document, the following terms are defined as follows.

The Association of National Park Authorities (ANPA)

Launched in 1996 as the successor to the Association of National Parks, ANPA brings together representatives from each national park authority (usually chairmen) to discuss matters of mutual concern.

ANPA meets other bodies and presents views on behalf of all of the national park authorities. Its objectives are: to represent the national park authorities of the UK at home and abroad and to provide a focus for shared experience and collaborative action. For more information contact www.anpa.gov.uk

Areas of Outstanding Natural Beauty (AONBs)

AONBs are designated under the National Parks and Access to the Countryside Act 1949. The primary purpose of designation is to conserve natural beauty.

Biodiversity Action Plans

Biodiversity includes all living things: the rich variety of species, habitats and the whole ecological systems that make up the living earth. Following the Rio Earth Summit in 1992, each country that signed up to the Convention on Biological Diversity must produce a national Biodiversity Action Plan setting out how it intends to conserve the living heritage and protect biological resources for the future. The UK plan was produced in 1994, and subsequent plans have been produced for Sussex and Hampshire.

Category V protected landscapes

National parks are part of a worldwide network of protected areas. In international terms they are classed as Category V protected landscapes by the World Conservation Union (IUCN). The management objectives for this category are "to maintain significant areas which are characteristic of the harmonious interaction of nature and culture, while providing opportunities for public enjoyment through recreation and tourism, and supporting the normal lifestyle and economic activity of these areas. These areas also serve scientific and educational purposes as well as maintaining biological and cultural diversity."

The Council for National Parks (CNP)

Provides a corporate voluntary sector voice on national park matters and acts as an umbrella organisation for a range of national bodies interested in national parks. CNP is composed of around 45 conservation and amenity organisations, all committed to national park objectives, and it co-opts a member from each national park authority.

CNP works to influence a wide range of policy and decision makers, oppose threats to national parks and create an informed constituency through the Friends of National Parks. For more information contact www.cnp.org.uk

Community plans/strategies

Part 1 of the Local Government Act 2000 places a duty on local authorities to prepare 'community strategies' (more commonly known as community plans) for the promotion or improvement of the economic, social and environmental well-being of their areas. The plans comprise four key components:
- a long-term vision for the area, focussing on the outcomes that are to be achieved;
- an action plan identifying short-term priorities and activities that will contribute to the achievement of long-term outcomes;
- a shared commitment to implement the action plan and proposals for doing so;
- arrangements for monitoring the implementation of the action plan and for periodically reviewing the community plan.

Countryside and Rights of Way Act 2000

The Countryside and Rights of Way Bill was introduced into Parliament in March 2000 and received Royal Assent on 30 November 2000.

The Act has five parts. Part I introduces new rights of access to open countryside. Part II introduces new rights of way legislation. Part III deals with nature conservation protection. Part IV introduces new powers to manage AONBs. Part V draws together miscellaneous and supplementary material.

Countryside Stewardship Scheme

This grant scheme operates throughout England outside of ESAs. Its aims are to:

- sustain the beauty and diversity of the landscape,
- improve and extend wildlife habitats,
- conserve archaeological sites and historic features,
- restore neglected land or features,
- create new habitats and landscapes,
- improve opportunities for people to enjoy the countryside.

This voluntary scheme is available to farmers and non-farming landowners and managers (including voluntary bodies and local authorities) who enter ten-year agreements to manage land in an environmentally beneficial way in return for annual payments. Grants are also available towards capital works.

Delegation/transfer

The practical differences between delegation and transfer relate to permanence and responsibility. Delegation is not necessarily permanent, as it can be reversed and ultimate responsibility remains with the authority that chooses to delegate. Transfer is permanent, usually by legislation, and moves complete responsibility to the authority to whom the powers have been transferred.

The Department for Environment, Food and Rural Affairs (DEFRA)

DEFRA has a crucial role in promoting sustainable development, rural renewal and sustainable and competitive food chains, both in the UK and internationally.

Their draft aim is: sustainable development and a better quality of life with:

- a better environment, diversity of wildlife and sustainable use of natural resources;
- economic prosperity through sustainable farming, fishing and food industries that meet consumers' requirements;
- thriving economies and communities in rural areas and a countryside for all to enjoy.

For more information, contact www.defra.gov.uk

The Department for Transport, Local Government and the Regions (DTLR)

DTLR aims to:

- promote modern and efficient local government in England, which is properly funded and responds to local needs;
- promote sustainable economic development in the English regions, including sponsorship of the RDAs and English Partnerships, and development of regional government policies for England;
- create a sustainable and integrated transport system, and provide the public with greater travel choice in making their journeys;
- achieve policies and provision for public transport users, pedestrians and motorists that are acceptable, accessible, available and affordable;
- be responsible for road policy, including national road scheme planning, maintenance, street works, traffic management, and monitoring traffic congestion.

For more information, contact www.dtlr.gov.uk

Designation Order

The secondary legislation which creates a national park and defines its boundary. Under the National Parks and Access to the Countryside Act 1949, the Countryside Agency is given statutory responsibility for the designation of national parks and making the Designation Orders.

East Hampshire Joint Advisory Committee

The East Hampshire AONB covers an area of 386 sq km encompassing the western end of the Downs from Petersfield to Winchester, the western fringe of the Weald and the valleys of the Meon and Rother. The East Hampshire Joint Advisory Committee (JAC) consists of representatives from local and national government organisations and statutory agencies, as well as from landowner, conservation, amenity and recreational bodies.

The JAC works in partnership with other organisations to conserve, protect and enhance the landscape, to foster the social and economic well-being of communities within the AONB and to promote quiet, informal enjoyment of the area by the public.

English Heritage
English Heritage is the lead body for the protection and understanding of England's historic environment, providing a comprehensive national source of expertise, skills and funding.

Its role includes advising government on all aspects of the historic environment; giving grants to secure the preservation of historic buildings and archaeological sites and monuments; encouraging the imaginative re-use of the nation's historic buildings to aid regeneration; and promoting the widest possible access to the heritage through the management of more than 400 historic properties that belong to the nation. For more information, contact www.english-heritage.org.uk

English Nature
A government agency that champions the conservation of wildlife and natural features throughout England. It promotes the protection and understanding of our geological heritage and advises government on wildlife, with powers and duties to protect and enhance natural heritage. English Nature owns and manages many National Nature Reserves and has responsibility for Sites of Special Scientific Interest. It also has responsibility for Special Areas of Conservation and Special Protection Areas, and for Ramsar sites. For more information, contact www.english-nature.org.uk

English Rural Development Programme (ERDP)
Sets out how the Department for Environment, Food and Rural Affairs is using the Rural Development Regulation (the 'second pillar' of the Common Agricultural Policy) to protect and improve the countryside and to encourage sustainable enterprise and thriving rural communities. It provides a framework for the operation of ten measures (some already operating, some new):
- Rural Enterprise Scheme (new),
- Vocational Training Scheme (new),
- Energy Crops Scheme (new),
- Processing and Marketing Grants Scheme (new),
- Environmentally Sensitive Areas Scheme,
- Countryside Stewardship,
- Organic Farming Scheme,
- Woodland Grant Scheme,
- Farm Woodland Premium Scheme,
- Hill Farm Allowance.

See also ESAs and the Countryside Stewardship Scheme.

Environmentally Sensitive Area (ESAs) Scheme
This scheme was introduced in 1987 to encourage farmers to help protect those areas of our countryside where the landscape, wildlife or historic interest is of national importance.

There are 22 ESA schemes in England and Wales, including the South Downs. Under the scheme, farmers voluntarily enter into ten-year management agreements with the Department for Environment, Food and Rural Affairs, for which they receive an annual payment on each hectare of land under agreement. Agreement holders have to follow specific management practices designed to conserve and enhance the landscape, historic and wildlife value of their land.

Establishment Order
This is the secondary legislation that creates the national park authority. Under the Environment Act 1995 section 63, the Secretary of State responsible may, by order, establish a national park authority.

Europarc (FNNPE)

The Federation of Nature and National Parks of Europe was established in 1973 to promote the activities of organisations and individuals concerned with nature parks and national parks at a European level. It encourages the designation of protected areas and the exchange of experience, training and mutual support for its members through staff exchanges, seminars, publications and working sessions.

The International Union for Conservation of Nature and Natural Resources (IUCN)

Increasingly known as the World Conservation Union, this is the largest professional body in the world that is working to care for the soil, land, water and air of our planet and the life that they support.

The IUCN is active in more than 120 countries, and includes 121 government agencies and more than 400 non-government conservation organisations, such as the World Wide Fund for Nature. It has categorised protected areas on a world scale and the national parks of England and Wales fall into Category V. It works in part through its own World Commission for Protected Areas. It has a UK Committee to which the Association of National Park Authorities belongs.

Local access forum

The Countryside and Rights of Way Act 2000 requires highway and national park authorities to set up local access forums. These statutory advisory bodies will advise on improving public access to land in their area for all types of open air recreation.

The views of local access forums must be taken into account when decisions are being taken on maps of open country, long-term restrictions on access land, appointing wardens, making byelaws and rights of way improvement plans. Forum members will include representatives of landowners and managers, recreational user groups (such as walkers and riders) and other local interests.

Local plans

These are plans, produced by district councils, some unitary authorities and national park authorities, in which policies are set out to guide development in the plan area. The local plan sets out detailed planning policies and specific proposals for development and use of land, and guide planning decisions. They must conform to structure plans.

The preparation of a local plan gives communities the opportunity to participate in planning choices about where development should be accommodated in their area. In providing the detailed framework for the control of development and use of land, local plans set out the authority's policies for this control. They also make proposals for the development and use of land and allocate land for specific purposes. Local plans cover the whole of a local authority's area.

Local transport plans (LTPs)

Following the White Paper, A new deal for transport; better for everyone (July 1998), all local authorities have been asked by the Government to prepare LTPs for their areas. These five-year integrated transport strategies, with statutory status, are the means by which local authorities plan and bid for resources for local transport initiatives.

Produced in consultation with local communities and partners, they include broad approaches towards widening transport choice, managing transport and the highway network, restraining demand, rural transport, and integration with wider policies.

Minerals and waste local plans

These are prepared by county planning authorities, non-metropolitan unitary authorities or national park authorities. Minerals and waste plans set out local authorities' detailed land use policies for the management and disposal of waste, within the broad strategic framework of the structure plan.

The minerals and waste local plan addresses the need for sites and facilities in particular areas, suitable locations and planning criteria likely to apply, including geological, hydrological and other considerations. They also carry forward policies which provide for the supply of minerals and for ensuring the required degree of environmental protection associated with development. They indicate, in more detail than structure plans, those areas where provision is made for mineral working and the disposal of mineral wastes and those areas where mineral resources are to be safeguarded for future working. They can also set out the development control criteria that will be applied in considering applications for mineral working and requirements for the restoration and aftercare of such sites.

The Ministry of Defence

The purpose of the Ministry of Defence, and the Armed Forces, is to defend the UK, and Overseas Territories, its people and interests and act as a force for good by strengthening international peace and security.

National park management plan

The Environment Act 1995 requires each national park authority to prepare and publish a national park management plan. The plans are intended to guide both the national park authorities themselves in carrying out all their functions and all other bodies and individuals who have an interest in the management of the park.

The process of preparing the plan is as important as the final plan. The plans are subject to widespread consultation and are developed and implemented in partnership.

Planning policy guidance notes (PPGs)

PPGs set out the government's policies on different aspects of planning. Local planning authorities must take their context into account in preparing their development plans. The guidance may also be material to decisions on individual planning applications and appeals. PPG 7 (revised), The Countryside - Environmental Quality and Economic and Social and Development, provides guidance on land use planning in rural areas.

Regionally Important Geological and Geomorphological Sites (RIGS)

RIGS were established in 1990 by the Nature Conservancy Council and continue to be actively supported by English Nature and other countryside agencies. RIGS are important sites that underpin and complement SSSI coverage.

RIGS are selected by voluntary, local RIGS groups that are generally formed by unitary authority area in England. RIGS are selected on a local or regional basis according to the following national criteria:

- the value of the site for educational purposes in life-long learning,
- the value of a site for study by both amateur and professional earth scientists,
- the historical value of a site from an earth science perspective,
- the aesthetic value of a site from an earth science perspective.

RIGS do not have statutory protection. However, they may be listed on local authorities' development plans. For more information, see www.english-nature.gov.uk

Relevant authorities

Defined as any Minister of the Crown, any public body, any statutory undertaker or any person holding public office.

Rights of way

The most widely known right to enjoy the countryside is that given by public rights of way. All public rights of way are highways in law. Anyone may use a public right of way, although it only gives 'a right of passage' to travel across the land and users must keep to the path. Rights of way include footpaths, bridleways and byways.

Rights of way improvement plan

Under the Countryside and Rights of Way Act 2000, highway authorities will have a duty to publish a rights of way improvement plan. The plan will be based on an assessment of how the rights of way network meets current and future needs of the public and should take account of the opportunities that local rights of way provide for exercise, recreation and enjoyment of the area. Detailed guidance on preparation of the plans, including consultation procedures, will be issued by the Minister responsible.

The South East England Development Agency (SEEDA)

SEEDA came into operation in 1999 to take the strategic lead in promoting the sustainable economic development of the region. Its mission is to work with partners to make the South East of England a world class region, achieving sustainable development and enhanced quality of life as measured by economic prosperity, environmental quality, social inclusion - ensuring meaningful employment for all. For more information, contact www.seeda.co.uk

The South East England Regional Assembly (SEERA)

SEERA was established in 1999 as the representative voice of the region. Its 111 members include elected councillors nominated by the region's local authorities. There are also regional representatives chosen by the voluntary sector, environmental groups, faith communities, business and economic partnerships, education and cultural networks and town and parish councils.

SEERA aims to be a strong and credible voice for the region, influencing the policies and programmes of central government, the European Commission and others in favour of the South East's interests. The Assembly has responsibility for proposing strategic planning and transport policies to government. For more information, contact www.southeast-ra.gov.uk

Site of Special Scientific Interest (SSSI)

A nationally important site for nature conservation designated under the Wildlife and Countryside Act 1981 (as amended).

Site of Nature Conservation Interest/ Site of Importance for Nature Conservation (SNCI/SINC)

A site that has been identified by local authorities as being important for wildlife. They either contain good examples of a particular type of semi-natural vegetation, or species of plants and animals that have a restricted distribution within the area, or are of significant nature conservation value within an urban area. SNCIs/SINCs are identified in structure and local plans and are usually protected through structure or local plan policies.

Sussex Downs Conservation Board

The Sussex Downs Conservation Board was established in 1992 as a national, six-year experiment. The Board has continued since then under an interim agreement with the local authorities and the Countryside Agency.

The Board works in partnership with other organisations to ensure that the qualities that make the Sussex Downs AONB special and distinctive are valued in any decision-making that may affect its character. The objectives of the Board are:

- to protect, conserve and enhance the natural beauty and amenity of the Sussex Downs AONB, including its physical, ecological and cultural landscape;
- to promote quiet informal enjoyment of the Sussex Downs AONB by the general public but only in so far as is consistent with the first objective;
- generally to promote sustainable forms of economic and social development, especially working with farmers and landowners to encourage land management which supports the two objectives above.

Special Area of Conservation (SAC)

Areas designated under the EC Directive on the Conservation of Natural Habitats and of Wild Fauna and Flora (The Habitats Directive) 1992 as being of European importance for habitats and species.

Structure plans

Structure plans are produced by county councils, some unitary authorities and national park authorities (in many cases on a joint basis). They provide the strategic policy framework for planning and development control locally; ensure that the provision for development is realistic and consistent with national and regional policy; and secure consistency between local plans from neighbouring areas (see local plans).

Draft boundary maps

Key to maps and sections

South Downs National Park –
draft boundary for public
consultation

Section A (part) &
Section W (part)

Draft boundary ———

Area outside draft
national park

0   1km

see Map 49

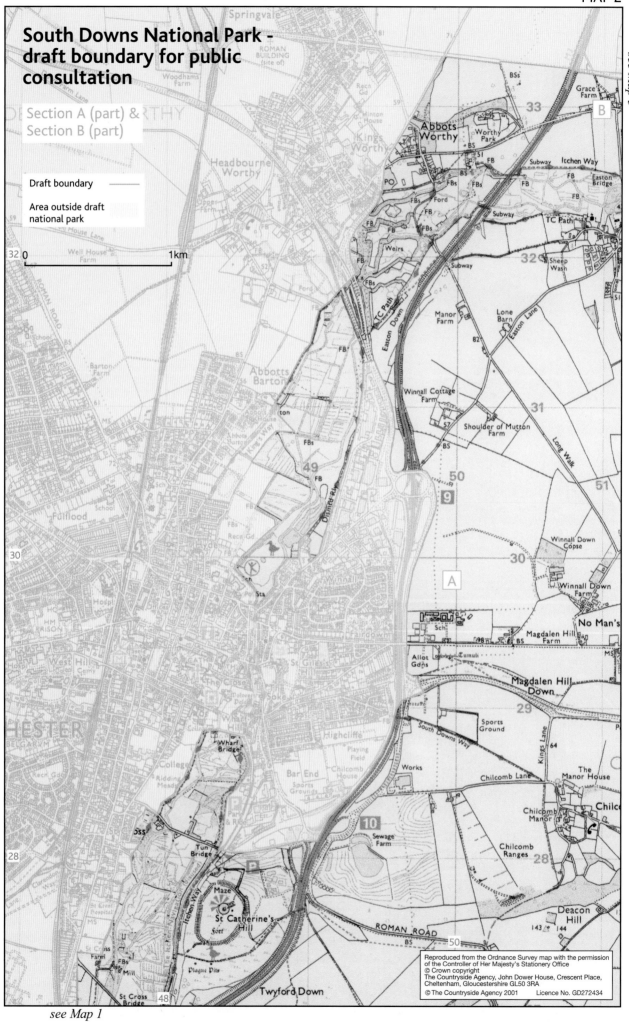

South Downs National Park - draft boundary for public consultation

Section A (part) & Section B (part)

Draft boundary

Area outside draft national park

0   1km

see Map 3

see Map 1

Reproduced from the Ordnance Survey map with the permission of the Controller of Her Majesty's Stationery Office
© Crown copyright
The Countryside Agency, John Dower House, Crescent Place, Cheltenham, Gloucestershire GL50 3RA
© The Countryside Agency 2001   Licence No. GD272434

MAP 3

see Map 4

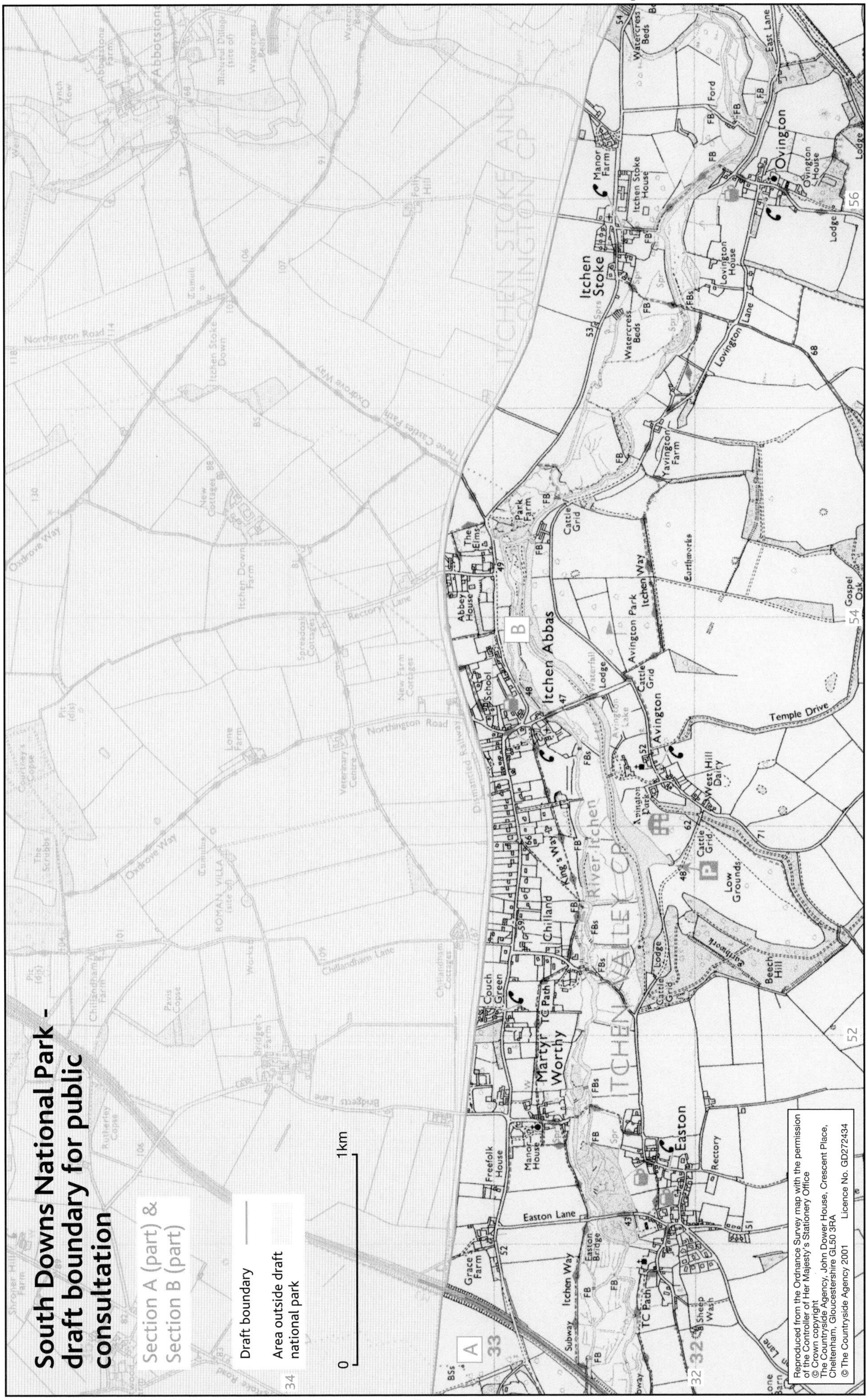

South Downs National Park – draft boundary for public consultation

Section A (part) & Section B (part)

Draft boundary

Area outside draft national park

0   1km

see Map 2

Itchen Stoke and Ovington CP
Ovington
Itchen Stoke
Manor Farm
Itchen Stoke House
Lovington House
Lovington Lane
Watercress Beds
Yavington Farm
Park Farm
The Elms
Abbey House
Itchen Abbas
School
Avington Park
Itchen Way
Avington Park Way
Earthworks
Temple Drive
Gospel Oak
West Hill Dairy
Avington
Avington Lake
Low Grounds
Beech Hill
River Itchen
Itchen Valley CP
King's Way
Chiland
Couch Green
Martyr Worthy
Easton
Rectory
Manor House
Freefolk House
Easton Lane
Easton Bridge
Grace Farm
Itchen Way
Sheep Wash
Abbotstone Farm
Bogacre Farm
Lynch Row
Northington Road
Itchen Stoke Down
Itchen Down Farm
Three Castles Path Oxdrove Way
Ovesdrove Way
Oxdrove Way
New Farm Cottages
Rectory Lane
Spreadeagle Cottages
Northington Road
Lone Farm
Roman Villa
Chilandham Lane
Chilandham Cottages
Bridges Farm
Augherley Copse
Courtleys Copse
Pavis Copse
The Scrubbs
Bridges Lane

A 33
B
32
34

56
52
97

see Map 5

MAP 4

South Downs National Park – draft boundary for public consultation

Section B (part) &
Section C (part)

—— Draft boundary

Area outside draft
national park

0   1km

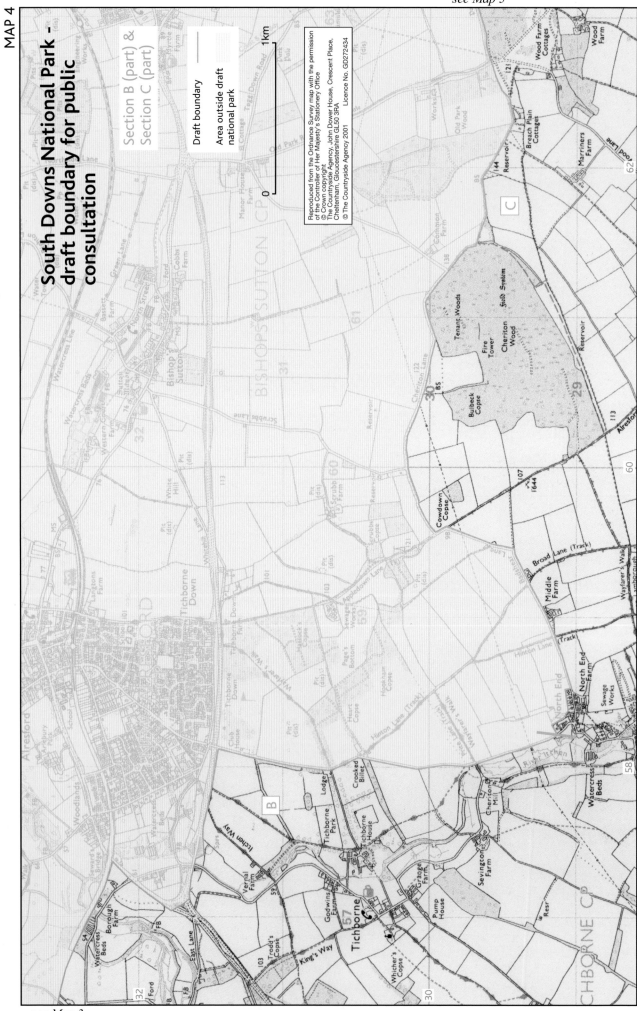

see Map 3

MAP 5

South Downs National Park – draft boundary for public consultation

Section C (part)

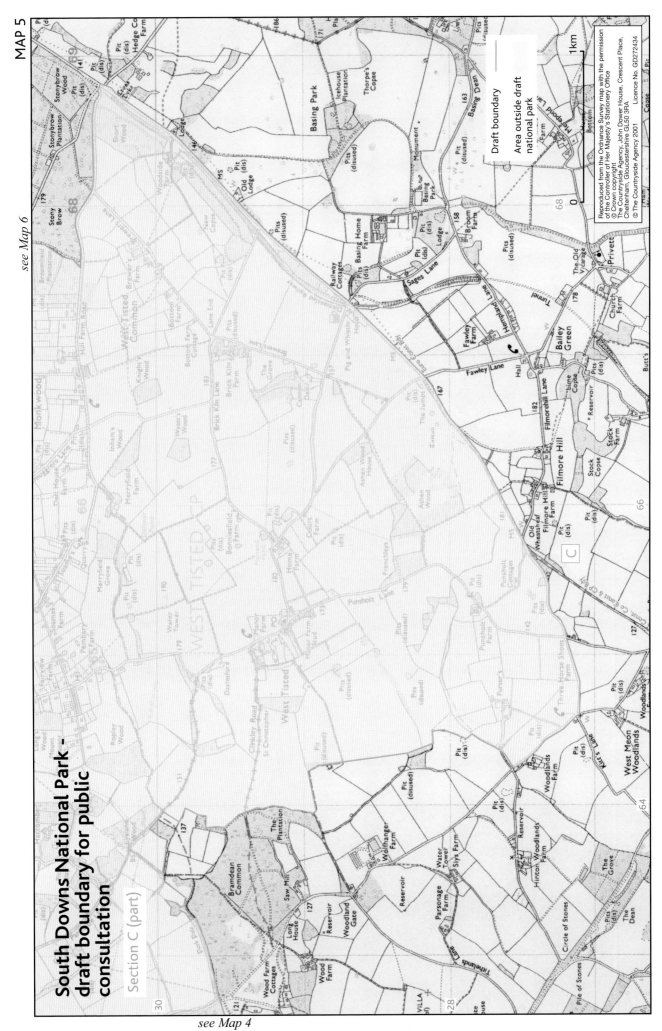

see Map 6

see Map 4

Draft boundary

Area outside draft national park

0   1km

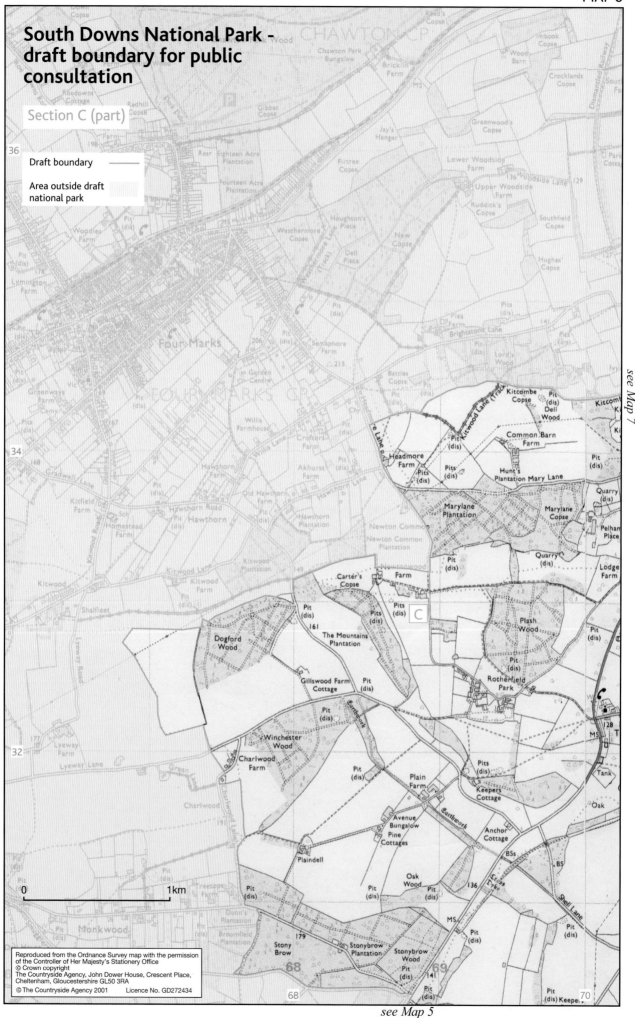

**South Downs National Park -
draft boundary for public
consultation**

Section C (part)

Draft boundary

Area outside draft
national park

see Map 7

see Map 5

MAP 7

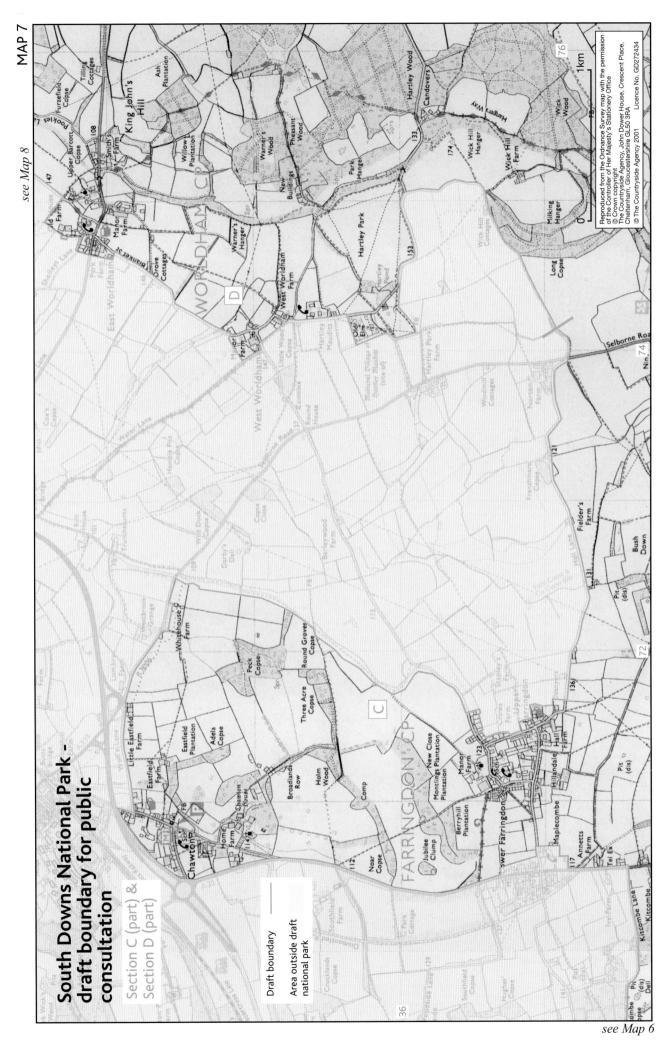

South Downs National Park – draft boundary for public consultation

Section C (part) & Section D (part)

Draft boundary

Area outside draft national park

see Map 8

see Map 6

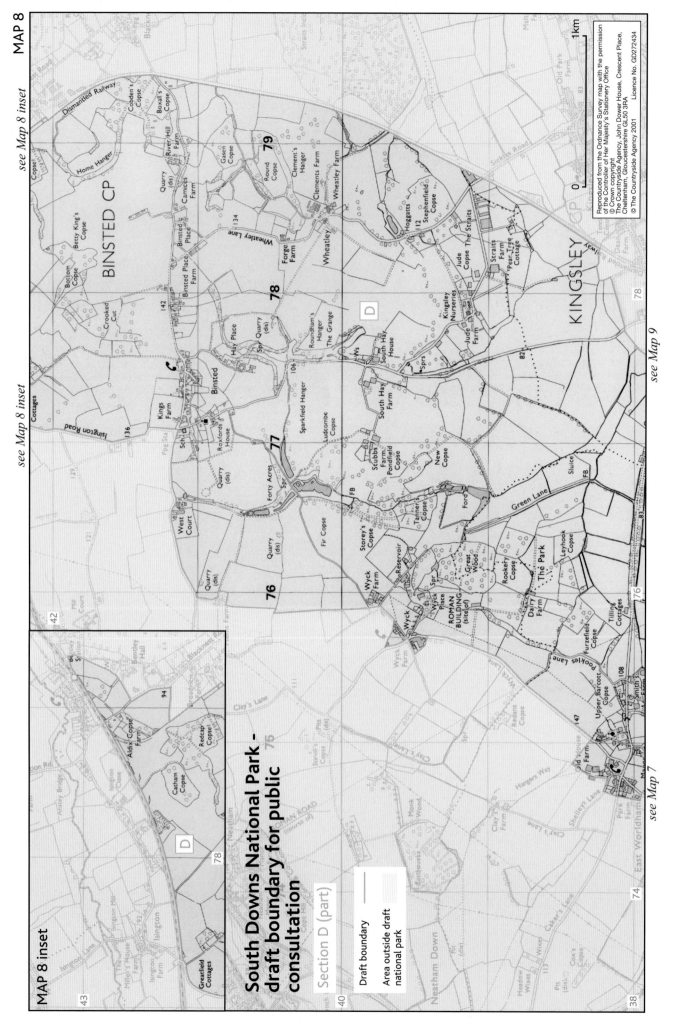

MAP 8

see Map 8 inset

MAP 8 inset

South Downs National Park – draft boundary for public consultation

Section D (part)

———— Draft boundary

Area outside draft national park

see Map 9

see Map 7

0   1km

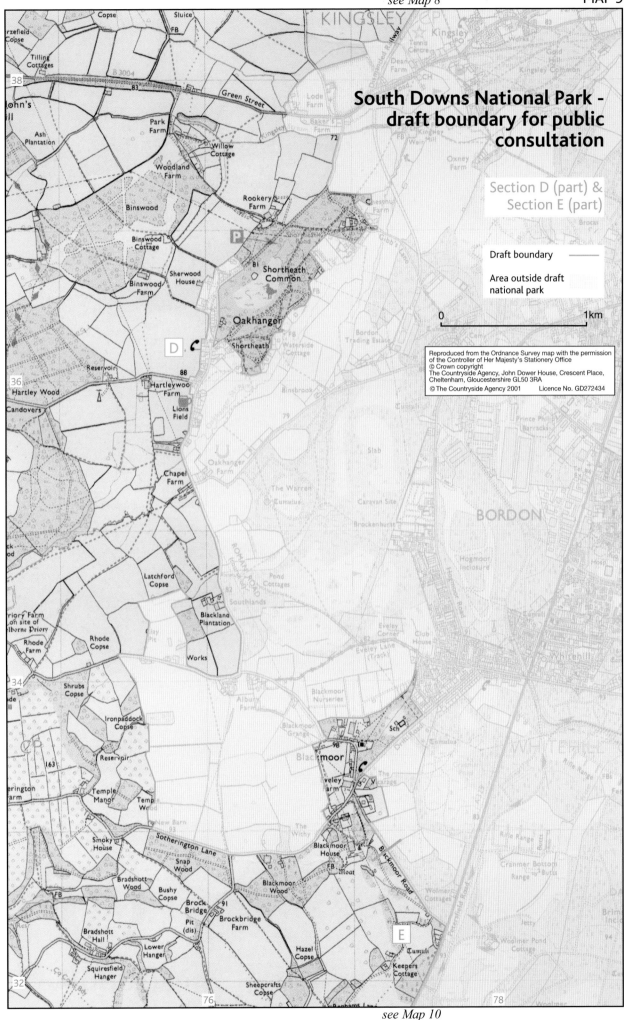

South Downs National Park -
draft boundary for public
consultation

Section D (part) &
Section E (part)

Draft boundary

Area outside draft
national park

0   1km

Reproduced from the Ordnance Survey map with the permission
of the Controller of Her Majesty's Stationery Office
© Crown copyright
The Countryside Agency, John Dower House, Crescent Place,
Cheltenham, Gloucestershire GL50 3RA

© The Countryside Agency 2001   Licence No. GD272434

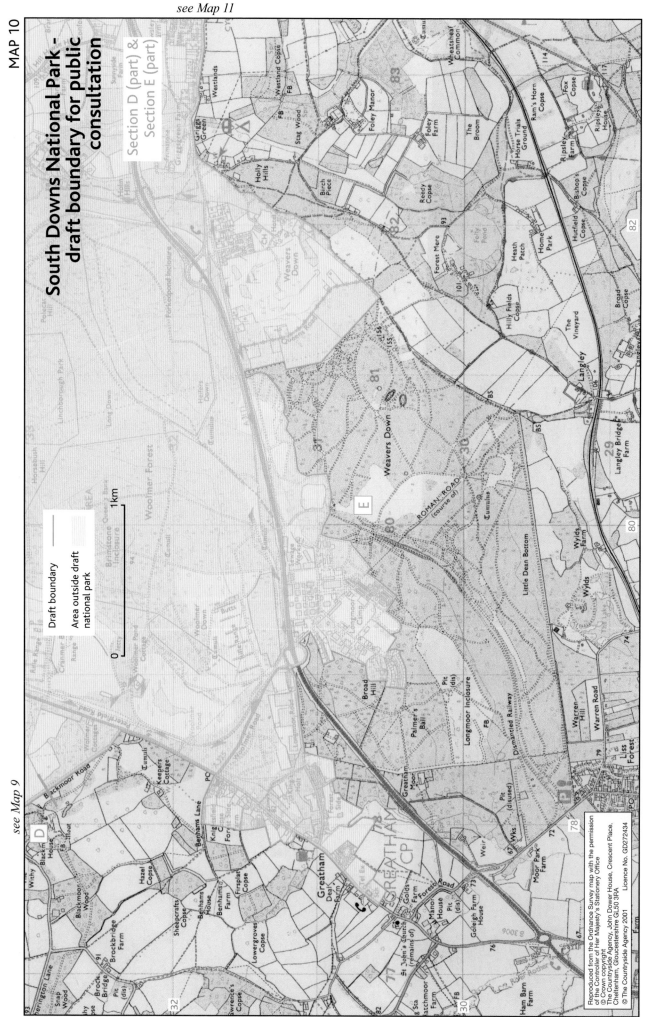

MAP 10

South Downs National Park –
draft boundary for public
consultation

Section D (part) &
Section E (part)

see Map 11

see Map 9

Draft boundary

Area outside draft
national park

0   1km

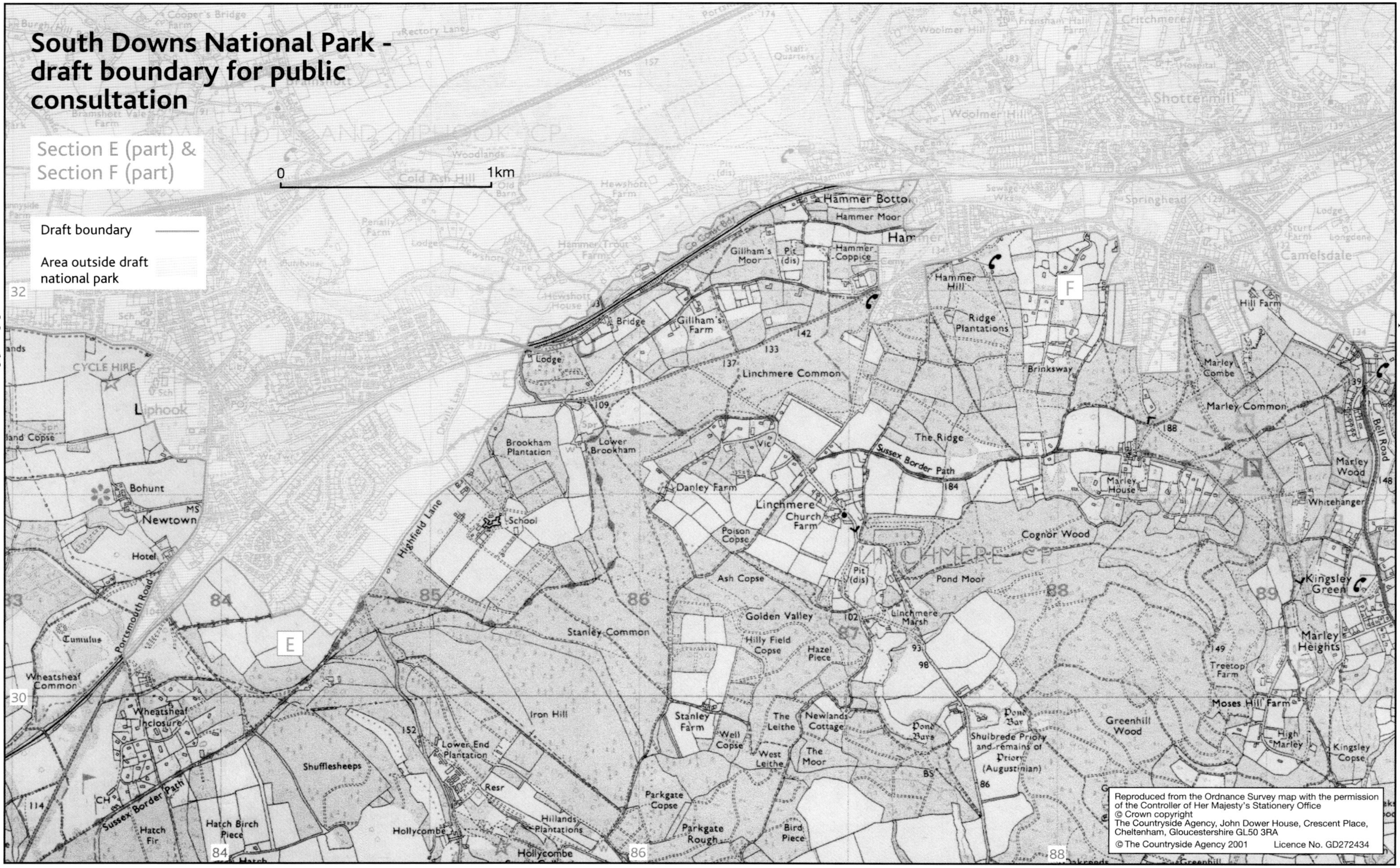

MAP 11

South Downs National Park – draft boundary for public consultation

Section E (part) & Section F (part)

Draft boundary

Area outside draft national park

see Map 12

see Map 10

0   1km

MAP 12

see Map 13

South Downs National Park – draft boundary for public consultation

Section F (part) & Section G (part)

Draft boundary

Area outside draft national park

0 ——— 1km

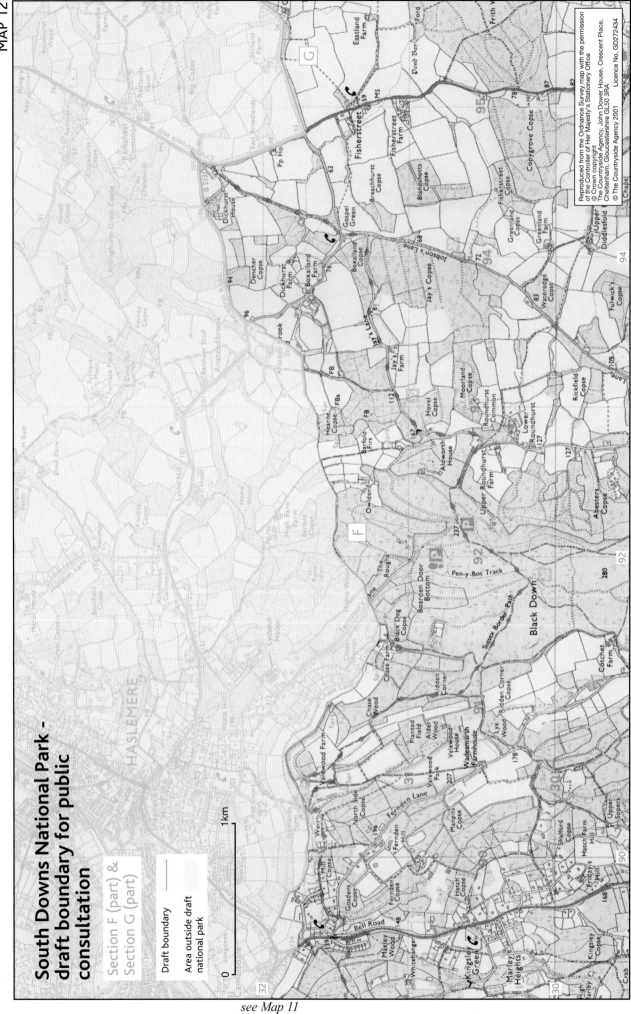

see Map 11

MAP 13

South Downs National Park – draft boundary for public consultation

Draft boundary

Area outside draft national park

0   1km

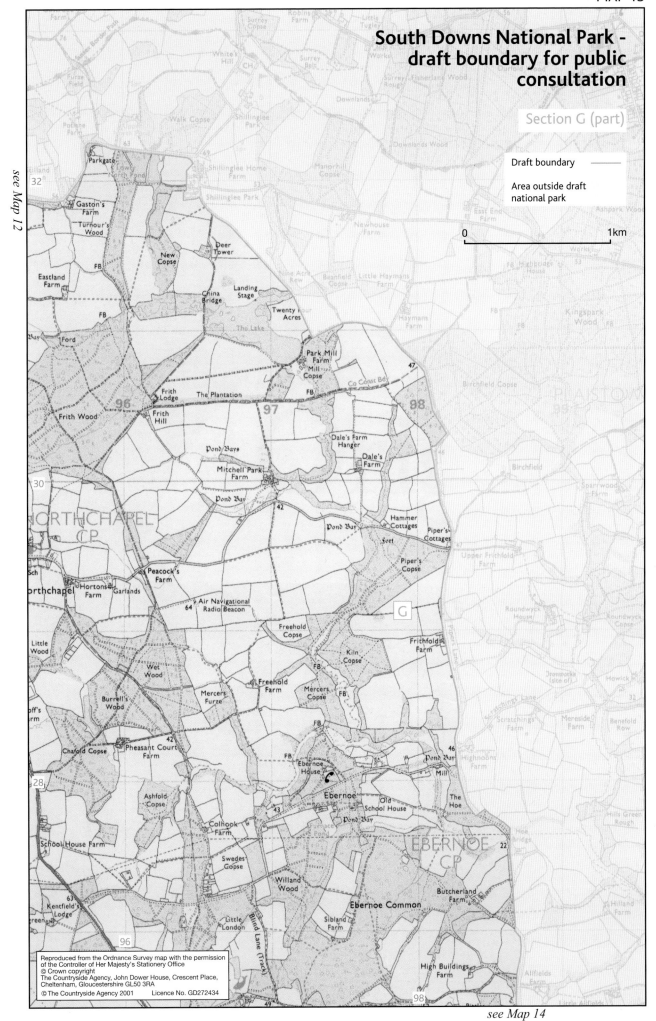

see Map 12

see Map 14

MAP 14

South Downs National Park - draft boundary for public consultation

Section G (part) &
Section H (part)

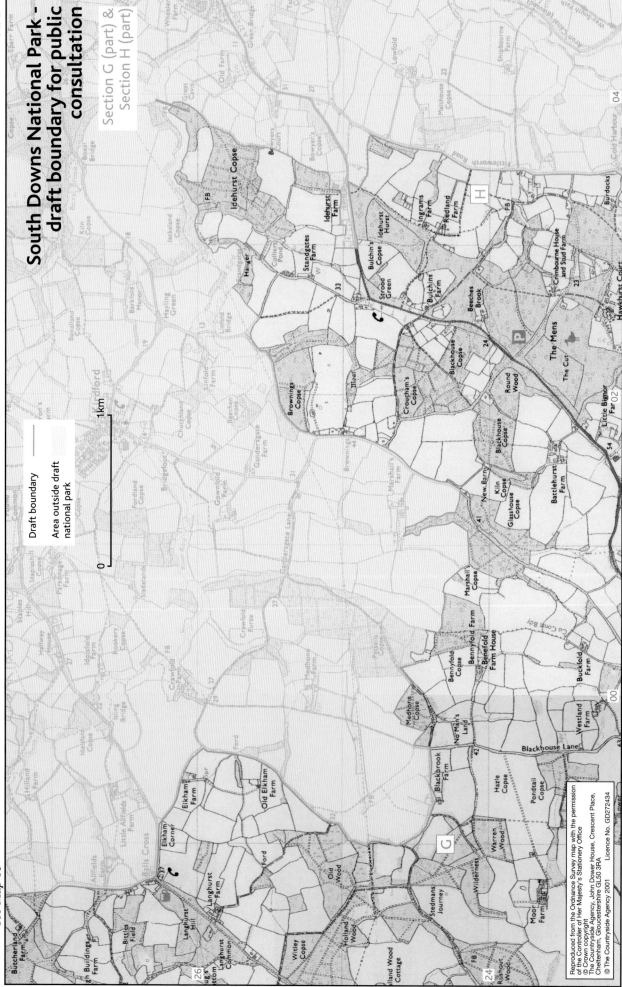

Draft boundary

Area outside draft
national park

0   1km

see Map 13

see Map 15

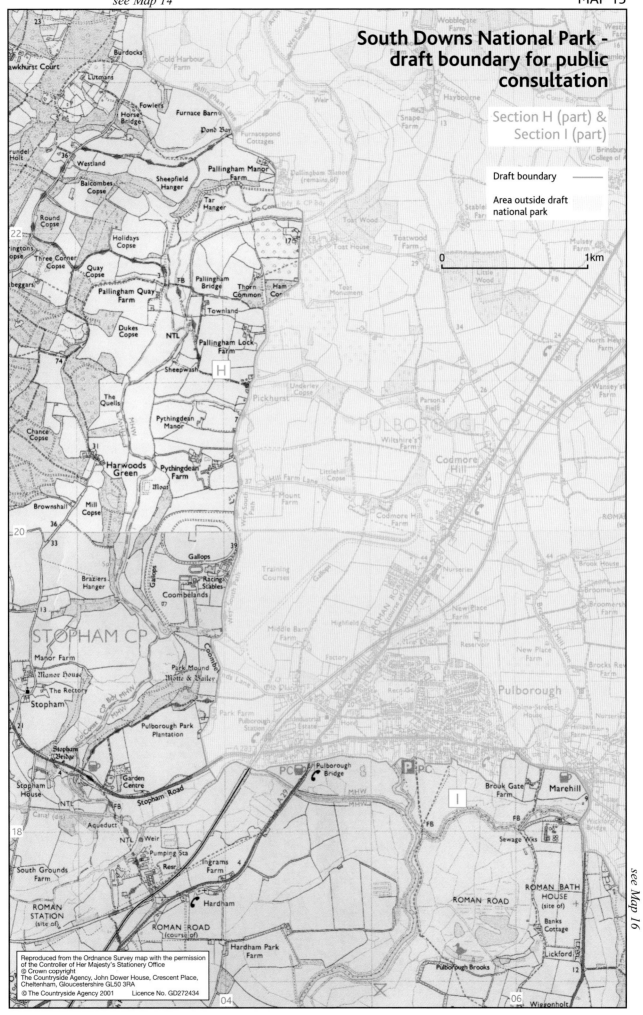

South Downs National Park -
draft boundary for public
consultation

Section H (part) &
Section I (part)

Draft boundary

Area outside draft
national park

0   1km

see Map 16

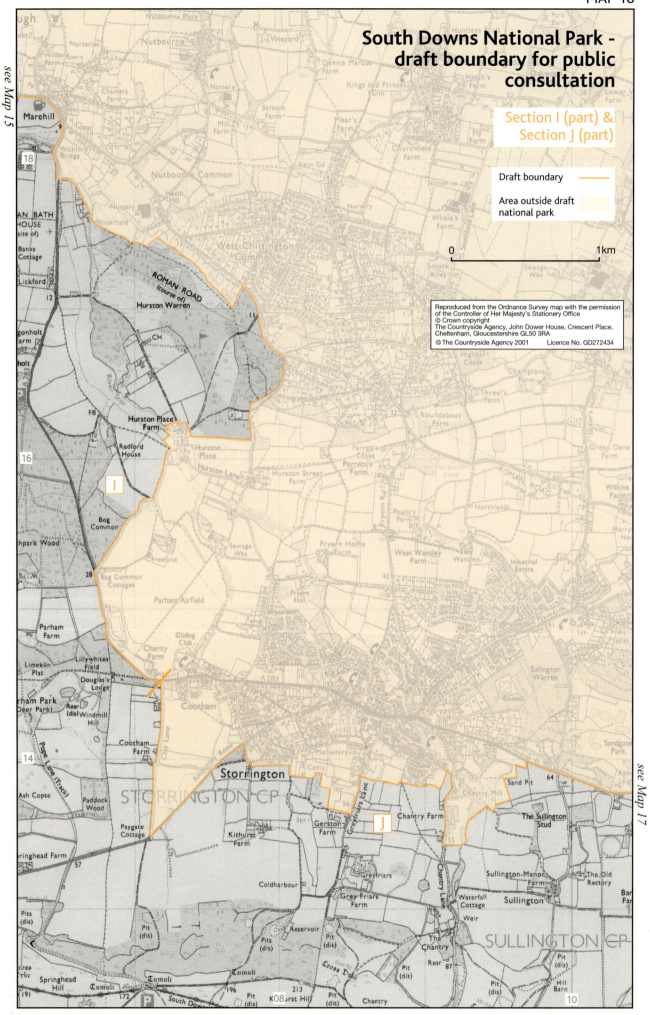

South Downs National Park - draft boundary for public consultation

Section I (part) &
Section J (part)

Draft boundary

Area outside draft
national park

0   1km

Reproduced from the Ordnance Survey map with the permission
of the Controller of Her Majesty's Stationery Office
© Crown copyright
The Countryside Agency, John Dower House, Crescent Place,
Cheltenham, Gloucestershire GL50 3RA
© The Countryside Agency 2001   Licence No. GD272434

MAP 17

see Map 18

South Downs National Park – draft boundary for public consultation

Section J (part) &
Section K (part)

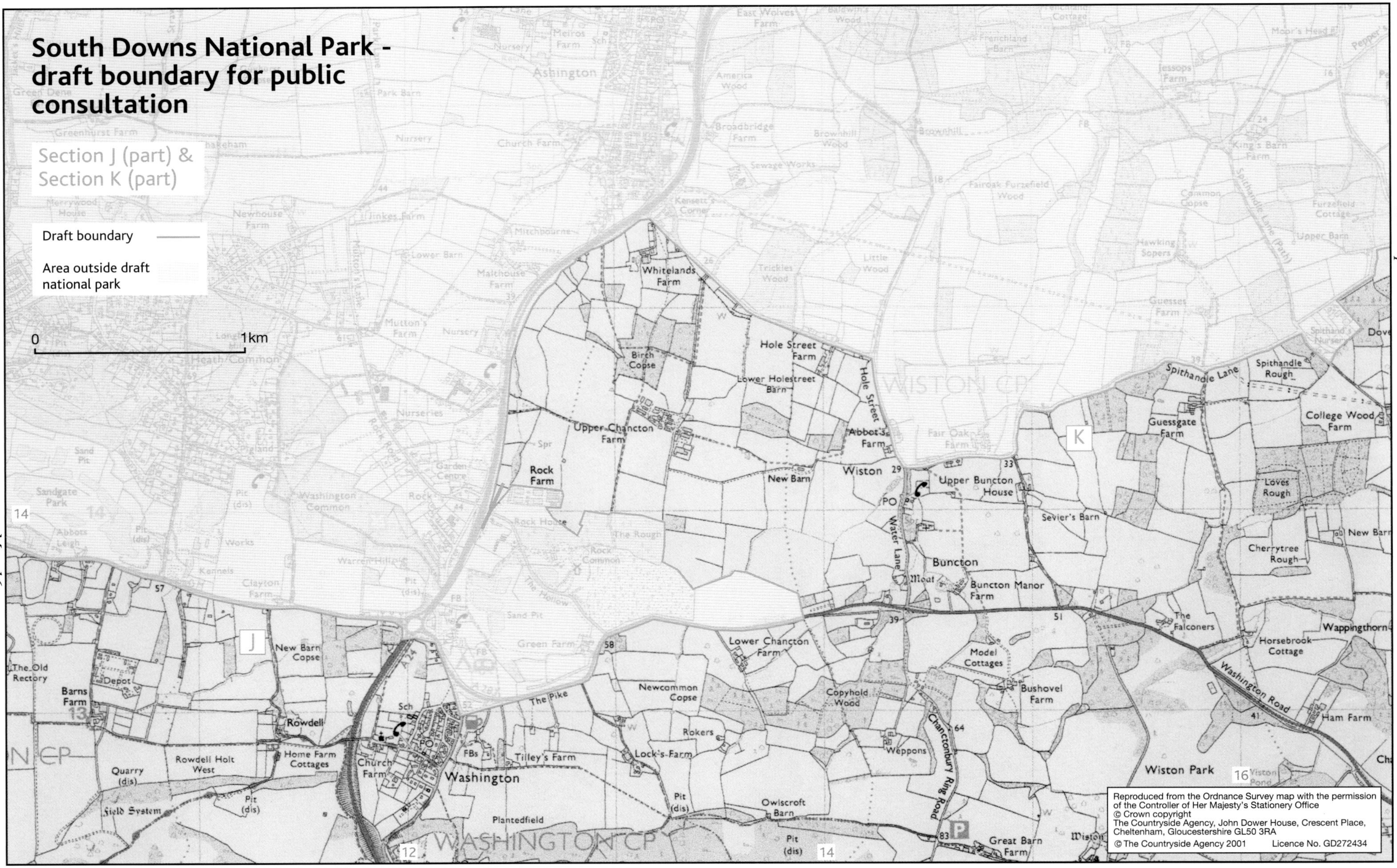

Draft boundary

Area outside draft
national park

0   1km

see Map 16

111

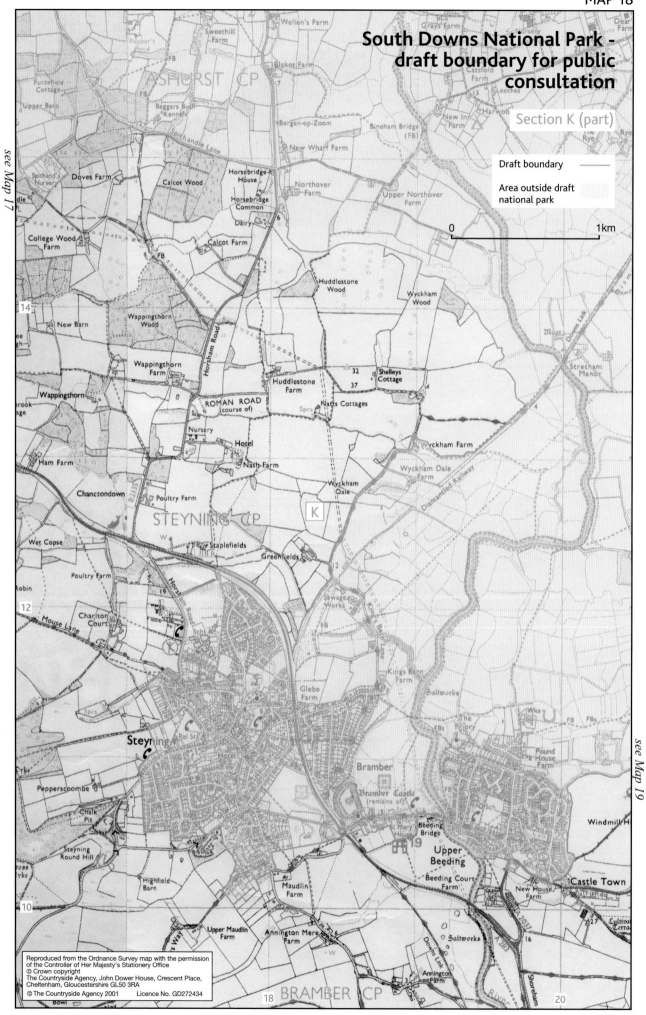

South Downs National Park - draft boundary for public consultation

Section K (part)

Draft boundary

Area outside draft national park

0   1km

MAP 19

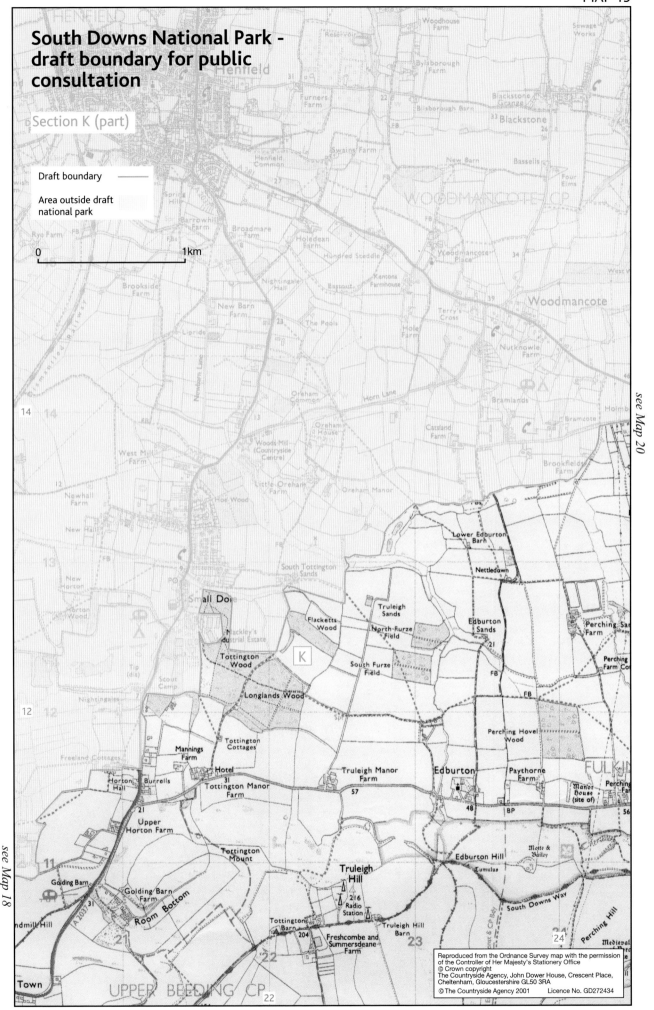

South Downs National Park - draft boundary for public consultation

Section K (part)

Draft boundary ————

Area outside draft national park

0   1km

see Map 20

see Map 18

MAP 20

see Map 21

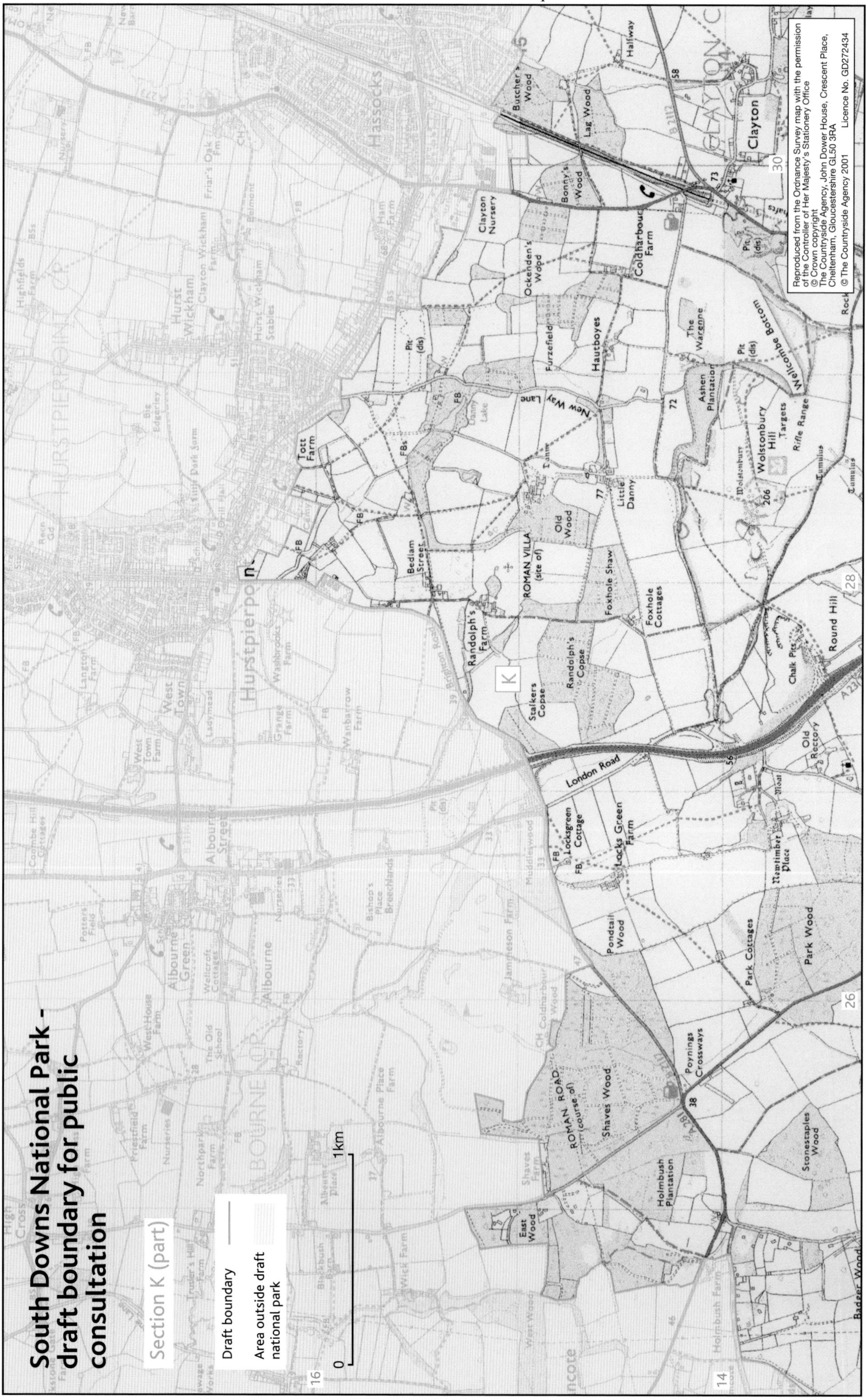

see Map 19

South Downs National Park - draft boundary for public consultation

Section K (part)

Draft boundary

Area outside draft national park

1km

0

MAP 21

see Map 22

South Downs National Park - draft boundary for public consultation

Section K (part)

Draft boundary ——

16 Area outside draft national park

0 ___ 1km

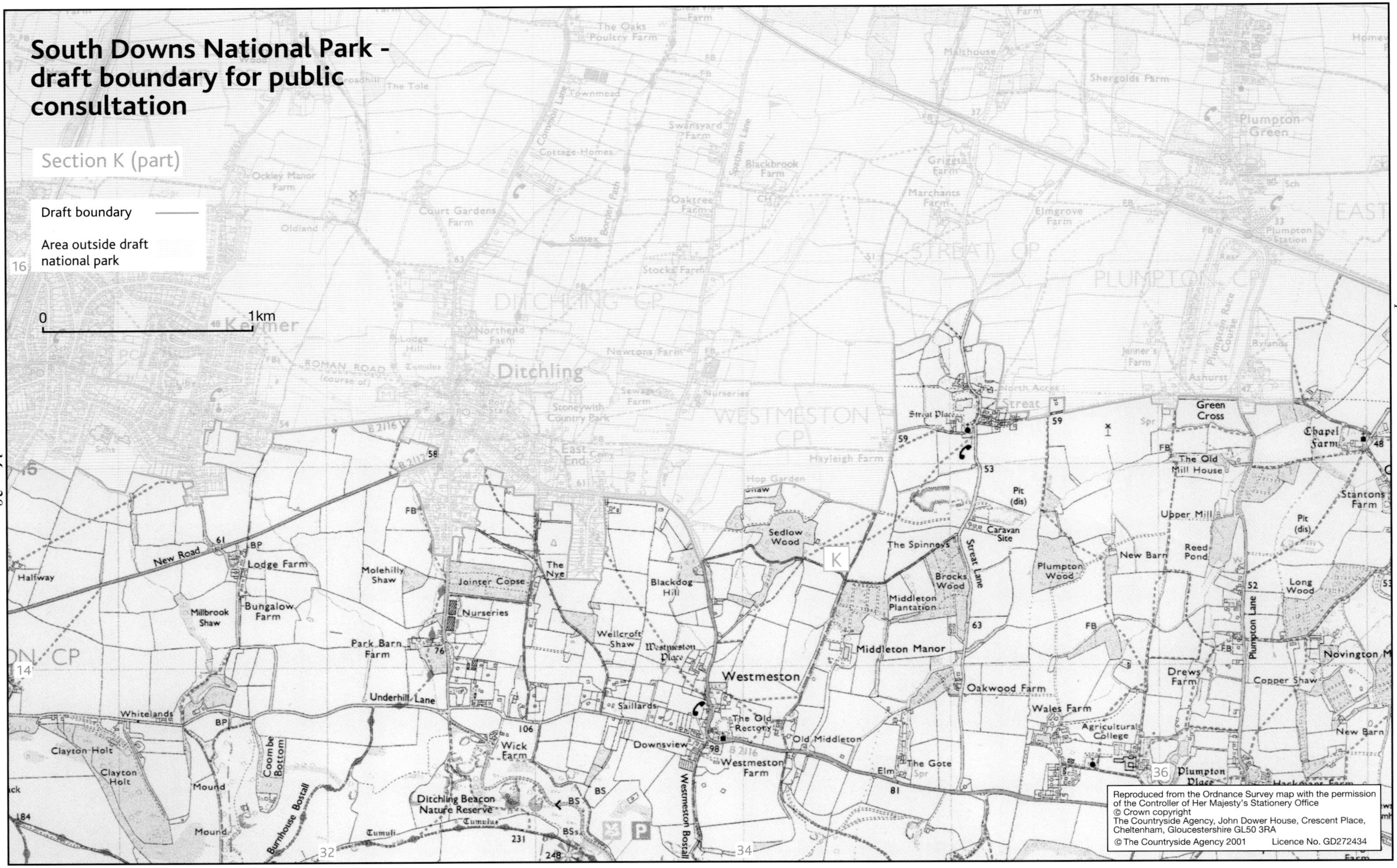

see Map 20

Plumpton Green
Sch
Plumpton Station
Plumpton Race Course
Ashurst
Shergolds Farm
Elmgrove Farm
Griggs Farm
Marchants Farm
Blackbrook Farm
Oakreed Farm
Stockel Farm
The Oaks Poultry Farm
Cottage Homes
Court Gardens Farm
Northend Farm
Newgates Farm
Muchouse Farm
Swansyard Farm
Ockey Manor Farm
Oldland
Border Path
Common Lane
Patcham Lane
Lodge Hill
Stoneywish Country Park
East End
Ditchling
DITCHLING CP
WESTMESTON CP
PLUMPTON
STREAT
Chapel Farm
Stantons Farm
Novington
New Barn
Copper Shaw
Long Wood
Pit (dis)
Plumpton Lane
Drews Farm
Plumpton Place
Agricultural College
Wales Farm
Oakwood Farm
Green Cross
The Old Mill House
Upper Mill
Reed Pond
New Barn
Plumpton Wood
FB
Pit (dis)
Caravan Site
Streat Lane
The Spinney
Brock Wood
Middleton Plantation
Middleton Manor
The Gote
Elm Spr
Old Middleton
Westmeston Farm
Westmeston
The Old Rectory
Westmeston Bostall
Downsview
Matthewson Place
Saillards
Blackdog Hill
Wellcroft Shaw
The Nye
Jointer Copse
Nurseries
Wick Farm
Underhill Lane
Park Barn Farm
Molehilly Shaw
Milbrook Shaw
Bungalow Farm
Lodge Farm
New Road
Whitelands
Clayton Holt
Ditchling Beacon Nature Reserve
Tumulus
Coombe Bottom
Burnhouse Bostall
Streat Place
Hayleigh Farm
Sedlow Wood
Hop Garden
231
248
106
76
81
63
53
59
52
36
34
32
98
48

MAP 22

South Downs National Park - draft boundary for public consultation

Section K (part) & Section L (part)

Draft boundary

Area outside draft national park

0   1km

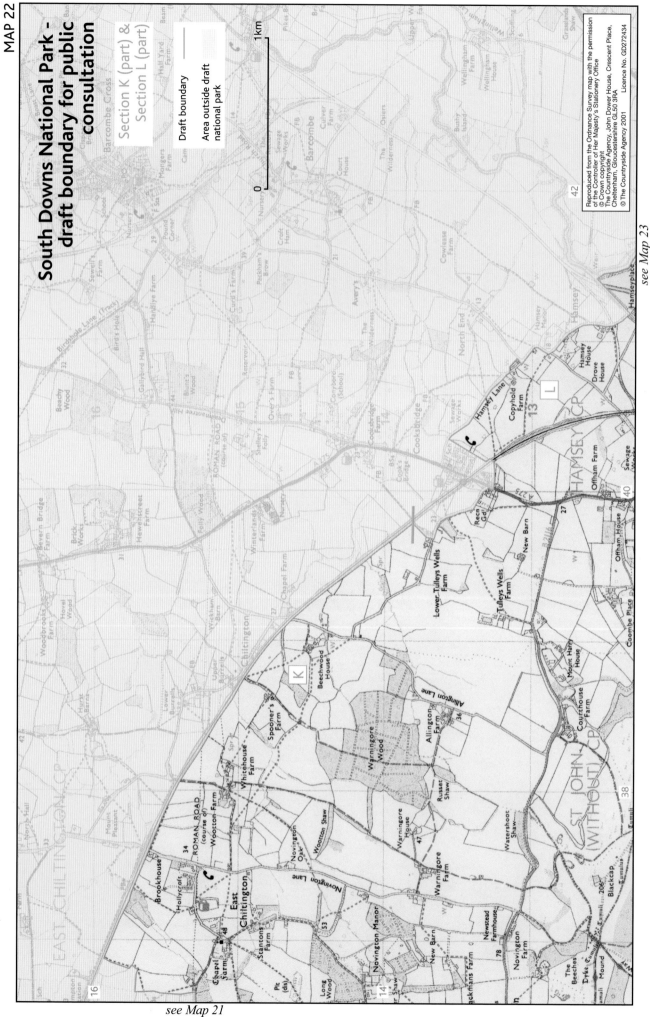

see Map 23

see Map 21

MAP 23

South Downs National Park - draft boundary for public consultation

Section L (part) & Section M (part)

Draft boundary

Area outside draft national park

0   1km

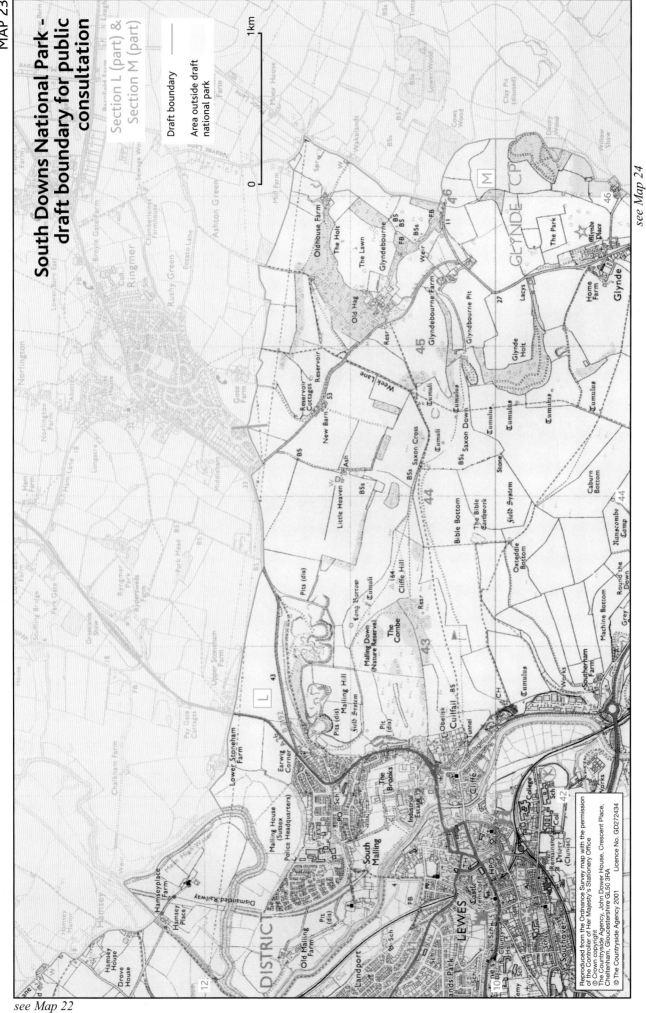

see Map 22

see Map 24

MAP 24

South Downs National Park – draft boundary for public consultation

Section M (part)

Draft boundary

Area outside draft national park

0   1km

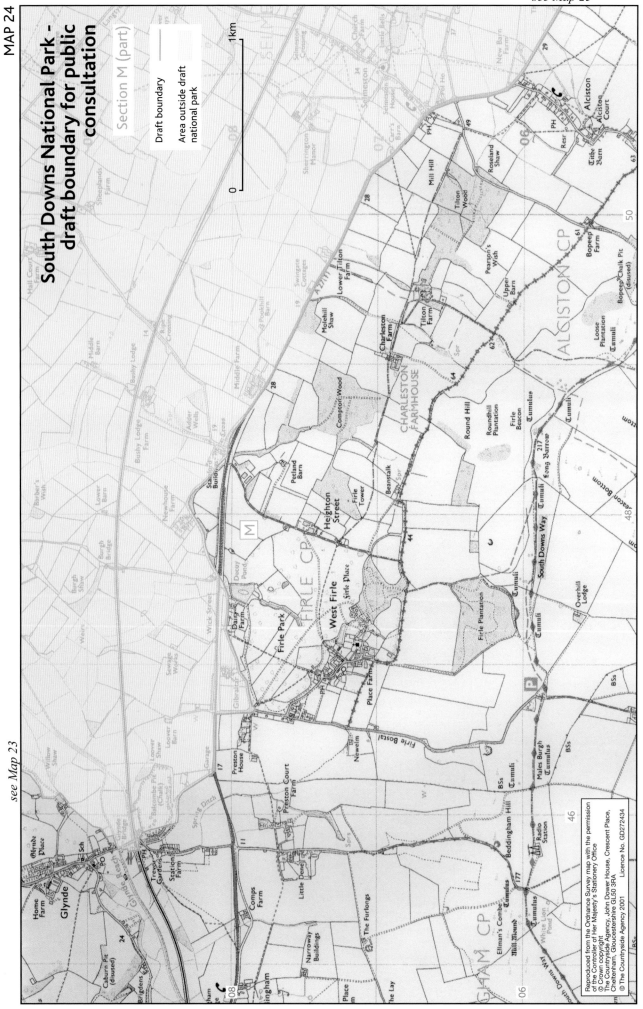

see Map 23

MAP 25

South Downs National Park - draft boundary for public consultation

Section M (part) & Section N (part)

see Map 26

see Map 24

Draft boundary

Area outside draft national park

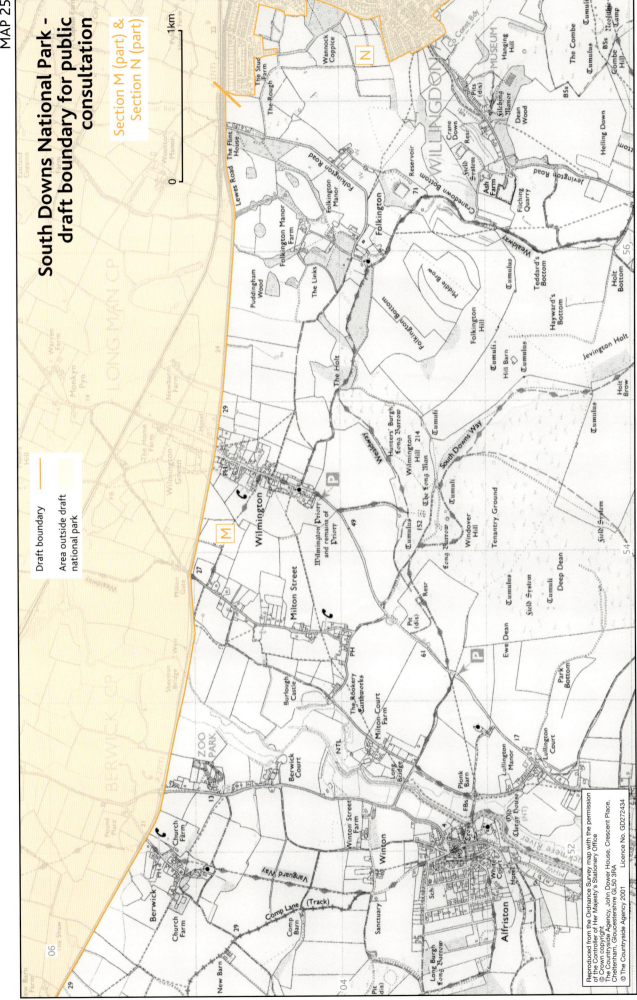

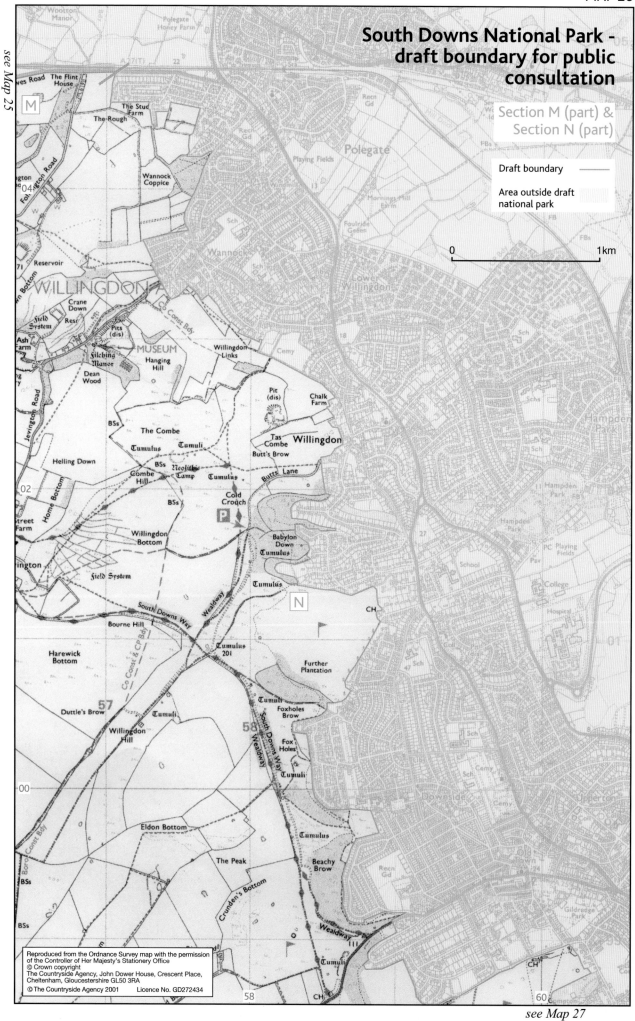

South Downs National Park – draft boundary for public consultation

Section M (part) & Section N (part)

Draft boundary

Area outside draft national park

0 _____ 1km

see Map 25

see Map 27

MAP 27

see Map 26

South Downs National Park – draft boundary for public consultation

Section N (part) & Section O (part)

—— Draft boundary

Area outside draft national park

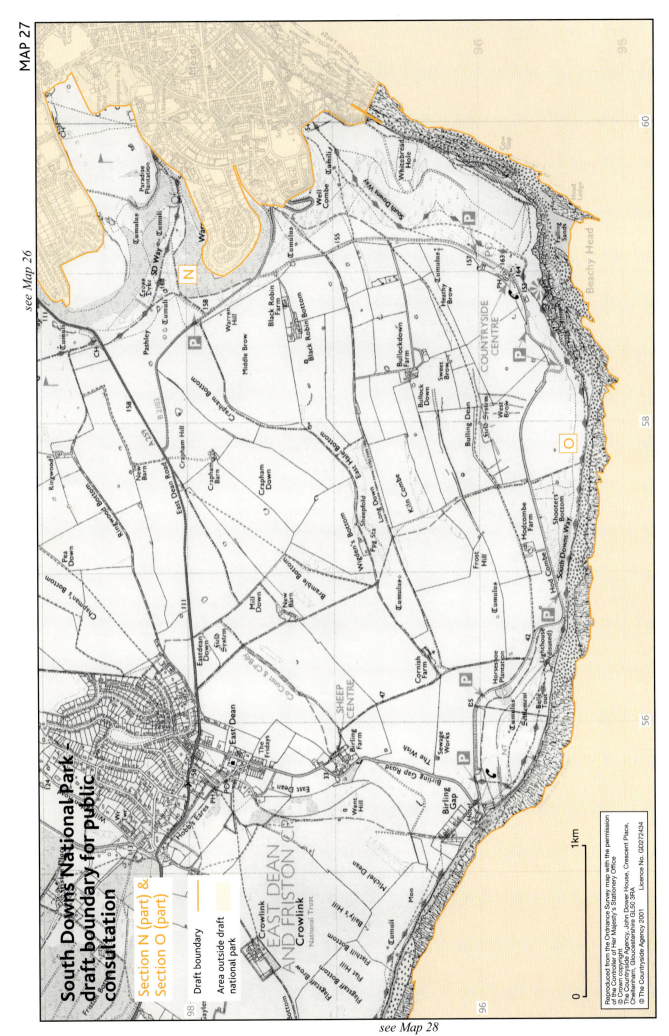

see Map 28

0   1km

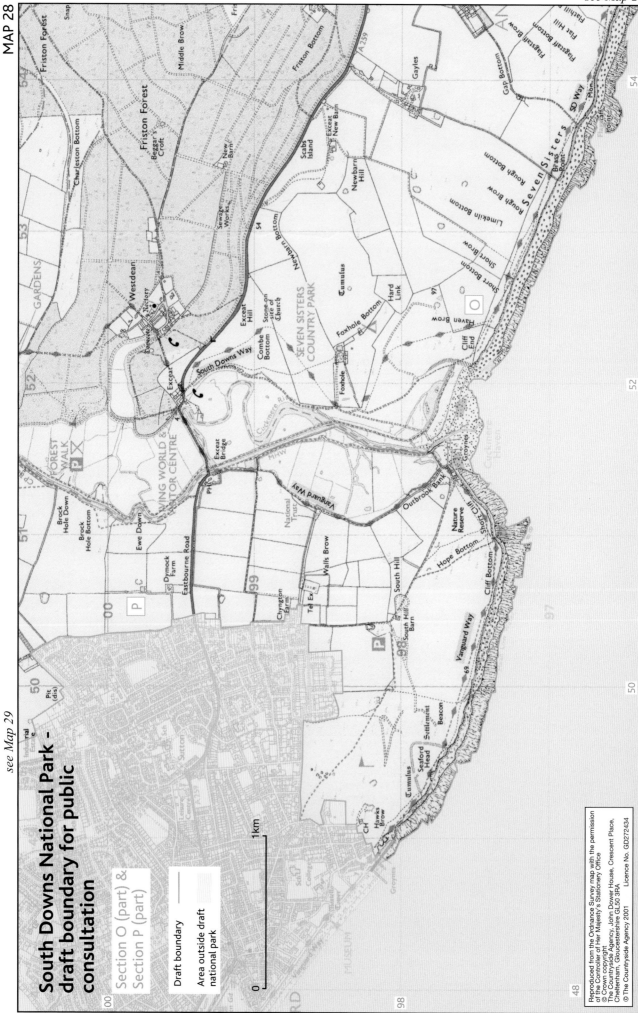

South Downs National Park -
draft boundary for public
consultation

Section O (part) &
Section P (part)

Draft boundary

Area outside draft
national park

1km

0

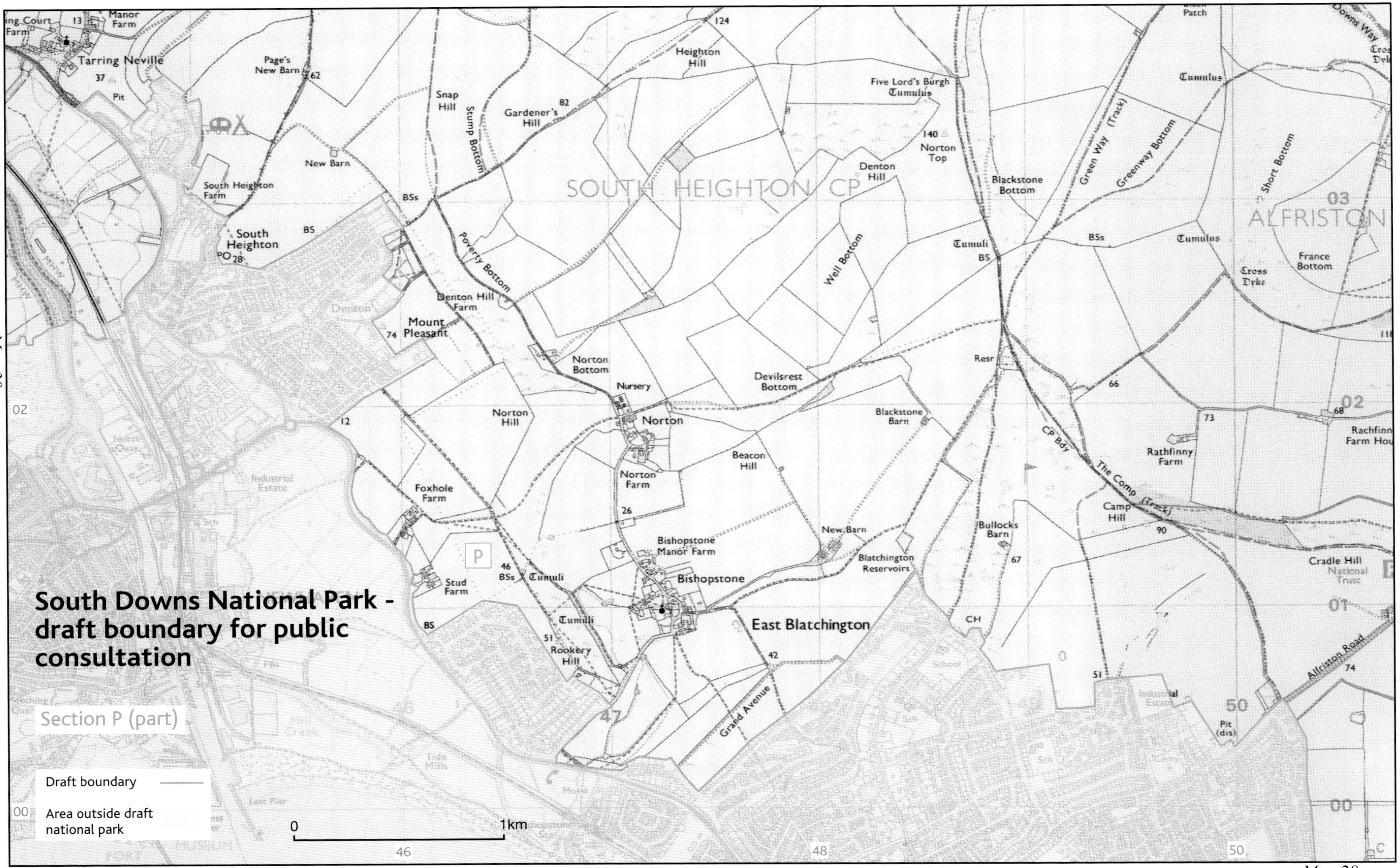

MAP 29

**South Downs National Park -
draft boundary for public
consultation**

Section P (part)

——— Draft boundary

Area outside draft
national park

0   1km

see Map 28

see Map 30

see Map 29

MAP 30

South Downs National Park -
draft boundary for public
consultation

Section P (part)

Draft boundary

Area outside draft
national park

Reproduced from the Ordnance Survey map with the permission
of the Controller of Her Majesty's Stationery Office
© Crown copyright
The Countryside Agency, John Dower House, Crescent Place,
Cheltenham, Gloucestershire GL50 3RA
© The Countryside Agency 2001   Licence No. GD272434

0   1km

see Map 31

MAP 31

see Map 30

see Map 32

see Map 32

see Map 32

see Map 32

Standean Bottom

Tumuli

The Bostle

Balsdean Reservoir

High Hill

Balsdean Cottages

Balsdean Farm

Pickers Farm

Nursery

Looes Barn

Rottingdean

New Barn

Rottingdean Place

Woodingdean Farm

Falmer Road

Recn Gd

B 2123

Resr

Ovingdean

Beacon Hill

St Dunstan's

Miniature Golf Course

Woodingdean

Happy Valley

Recn Gd

Schs

Hosp

Mount Pleasant

KEMPTOWN BC

Wick Bottom

P

P

Sanatorium

Castle Hill

Roedean

Roedean School

Red Hill

Roedean Bottom

BRIGHTON

Whitehawk

Sheepcote Valley

East Brighton Park

Black Rock

Race Course

Works

Kemp Town

Brighton Marina

Disused Railway

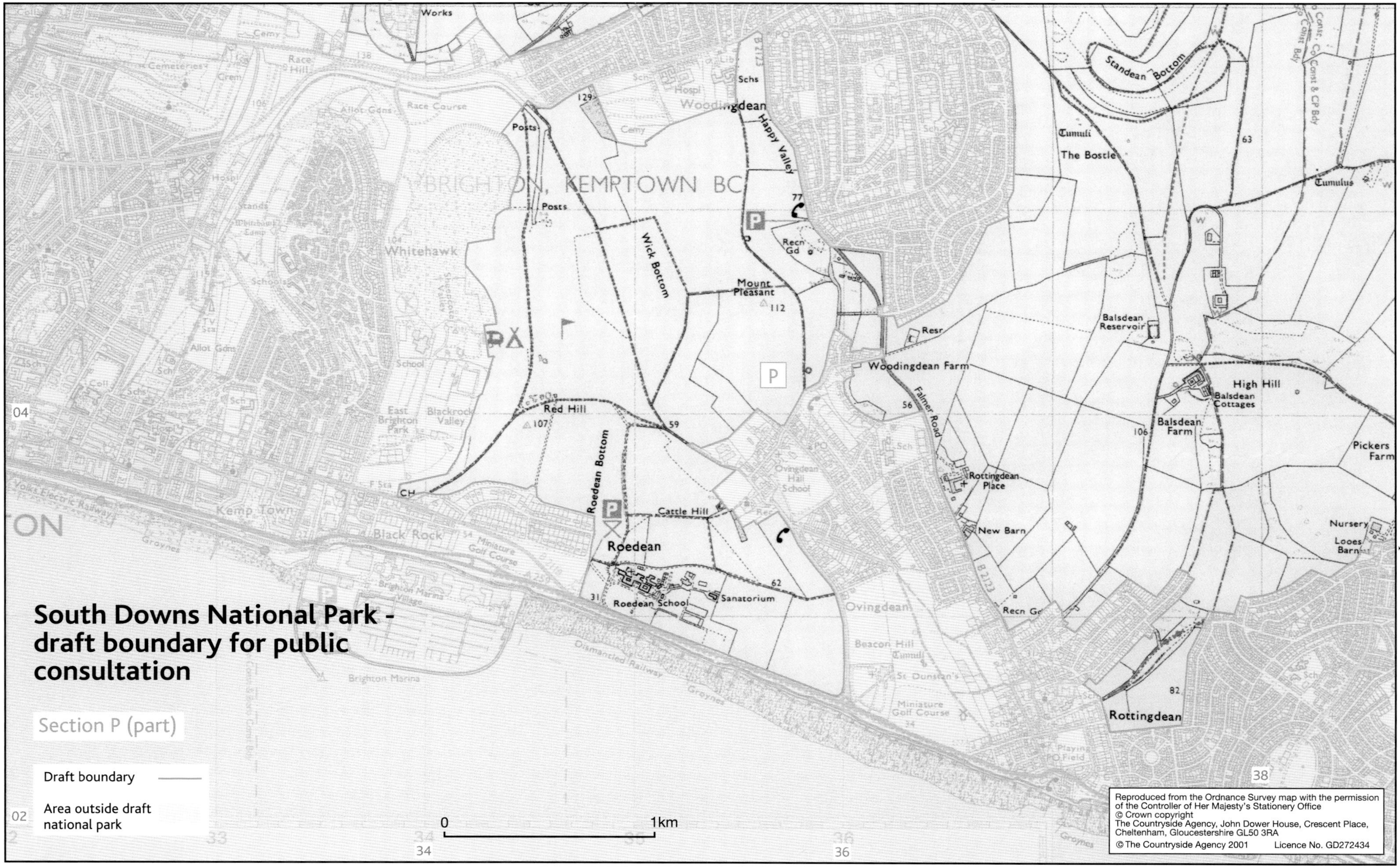

South Downs National Park - draft boundary for public consultation

Section P (part)

──────── Draft boundary

Area outside draft national park

0   1km

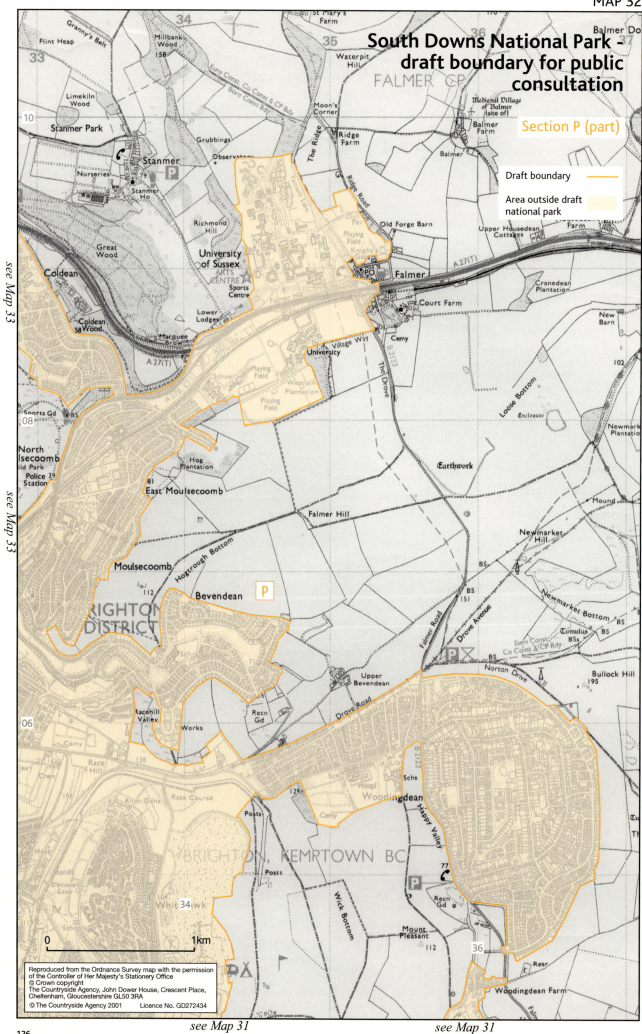

South Downs National Park -
draft boundary for public
consultation

Section P (part)

Draft boundary

Area outside draft
national park

see Map 31   *see Map 31*

MAP 33

South Downs National Park – draft boundary for public consultation

Section P (part)

see Map 32   see Map 32

see Map 32

see Map 34

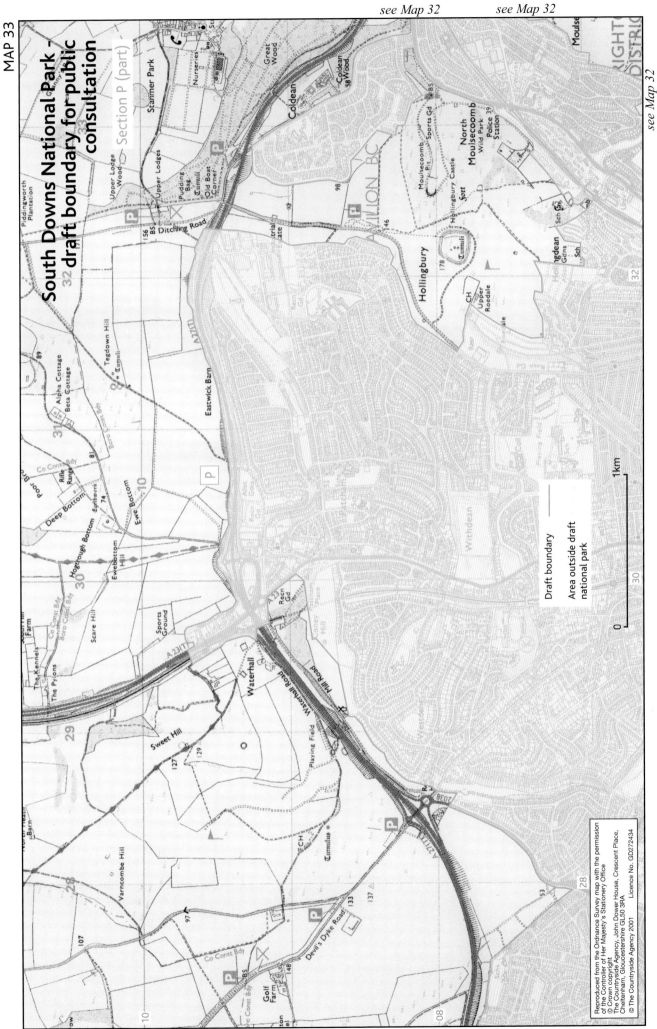

Draft boundary

Area outside draft
national park

0   30   1km

see Map 33

MAP 34

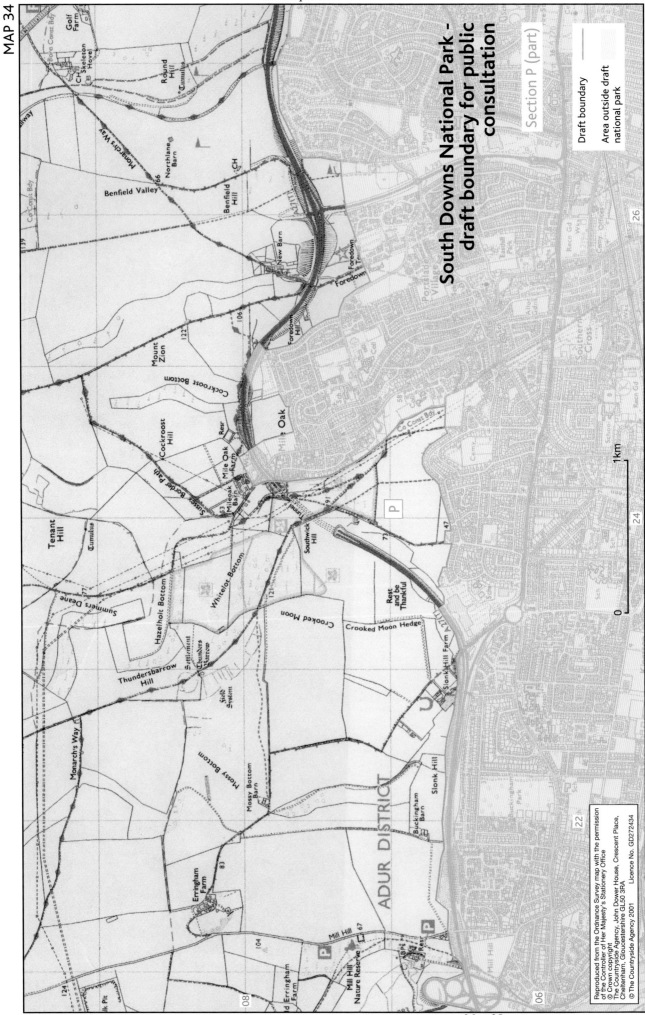

South Downs National Park - draft boundary for public consultation

Section P (part)

Draft boundary

Area outside draft national park

see Map 35

MAP 35

see Map 34

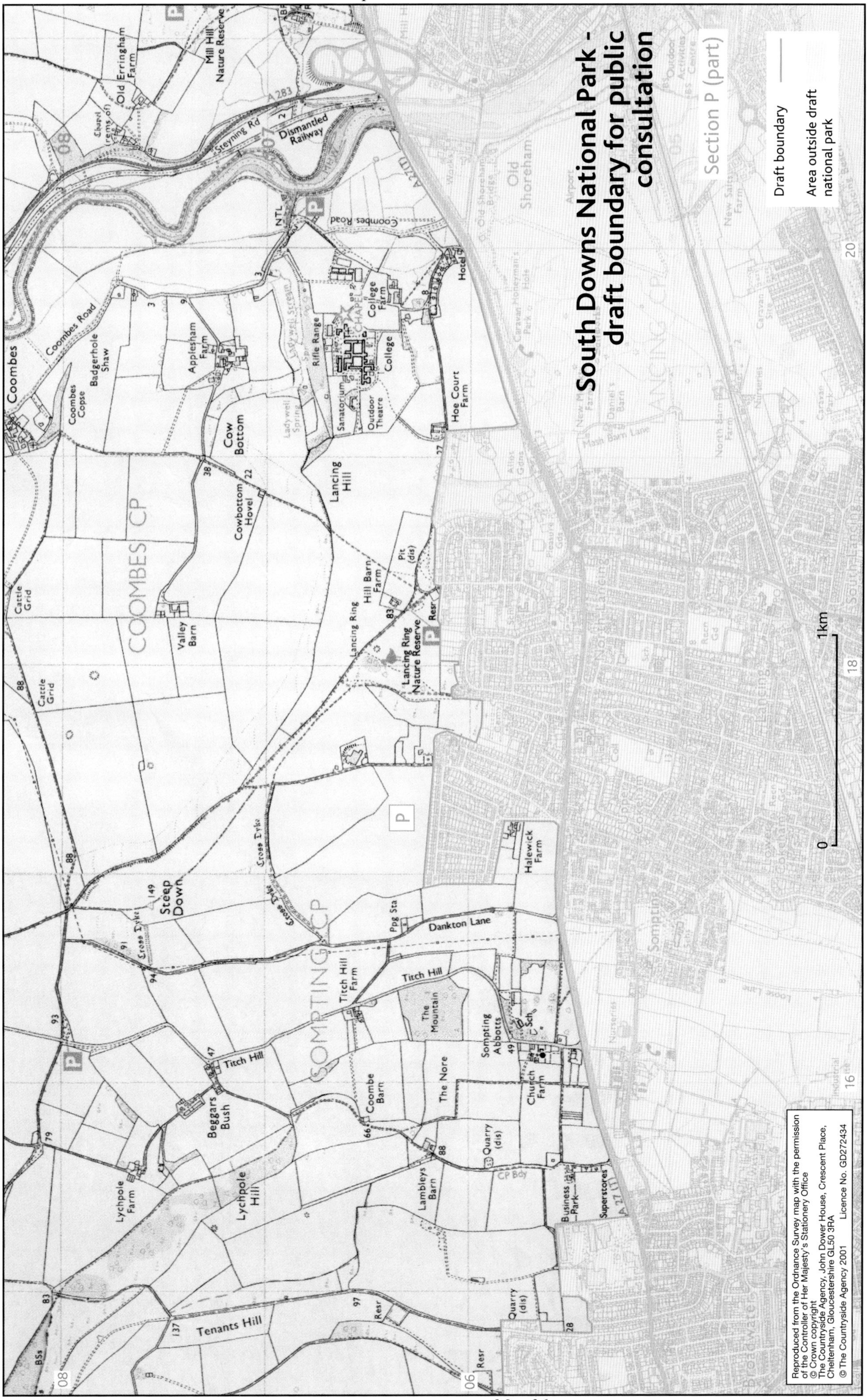

South Downs National Park – draft boundary for public consultation

Section P (part)

Draft boundary —

Area outside draft national park

Coombes

Coombes CP

COOMBES CP

Old Erringham Farm

Mill Hill Nature Reserve

Chapel (rems of)

Steyning Rd

Dismantled Railway

Coombes Road

Coombes Road

Coombes Copse

Badgerhole Shaw

Applesham Farm

Rifle Range

Sanatorium

CHAPEL

Outdoor Theatre

College

College Farm

Hoe Court Farm

Old Shoreham

Cow Bottom

Cowbottom Hovel

Lancing Hill

Hill Barn Farm

Pit (dis)

Lancing Ring

Lancing Ring Nature Reserve

Resr

Valley Barn

Cattle Grid

Cattle Grid

Lancing CP

North Lancing

New Salts Farm

Steep Down

Cross Dyke

Cross Dyke

Cross Dyke

SOMPTING CP

Ppg Sta

Dankton Lane

Halewick Farm

Sompting

Titch Hill Farm

Titch Hill

The Mountain

Sompting Abbotts

Beggars Bush

Titch Hill

Coombe Barn

The Nore

Church Farm

C Sch

Lychpole Farm

Lychpole Hill

Lambleys Barn

Quarry (dis)

Quarry (dis)

Business Park

Superstores

Tenants Hill

Resr

Resr

0   1km

see Map 36

MAP 36

see Map 35

South Downs National Park - draft boundary for public consultation

Section P (part) & Section Q (part)

Lodges

Draft boundary

Area outside draft national park

see Map 37

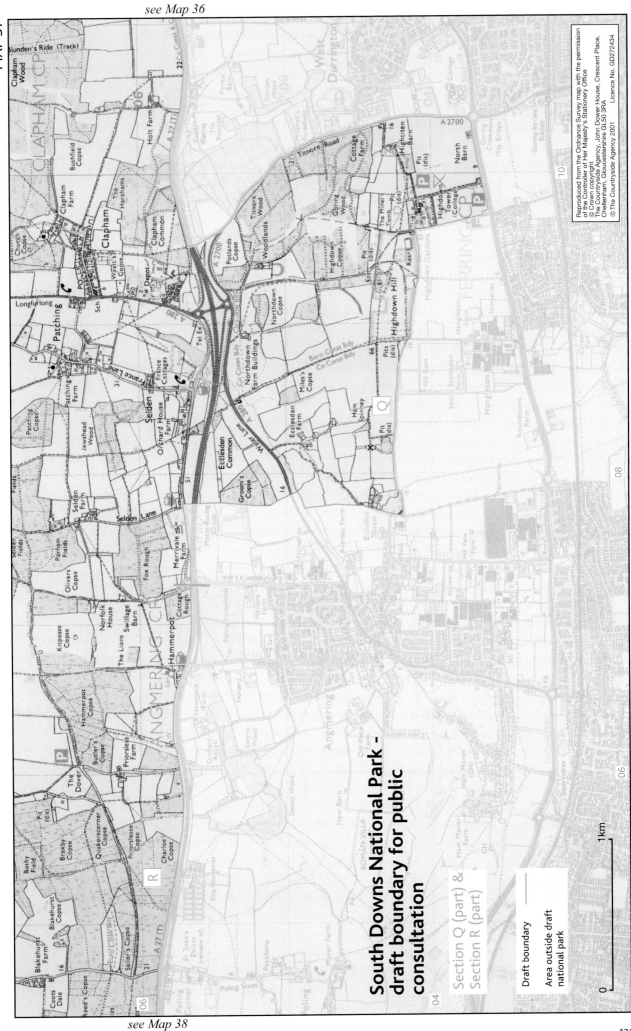

MAP 37

see Map 36

South Downs National Park – draft boundary for public consultation

Section Q (part) & Section R (part)

Draft boundary

Area outside draft national park

0 1km

MAP 38

see Map 37

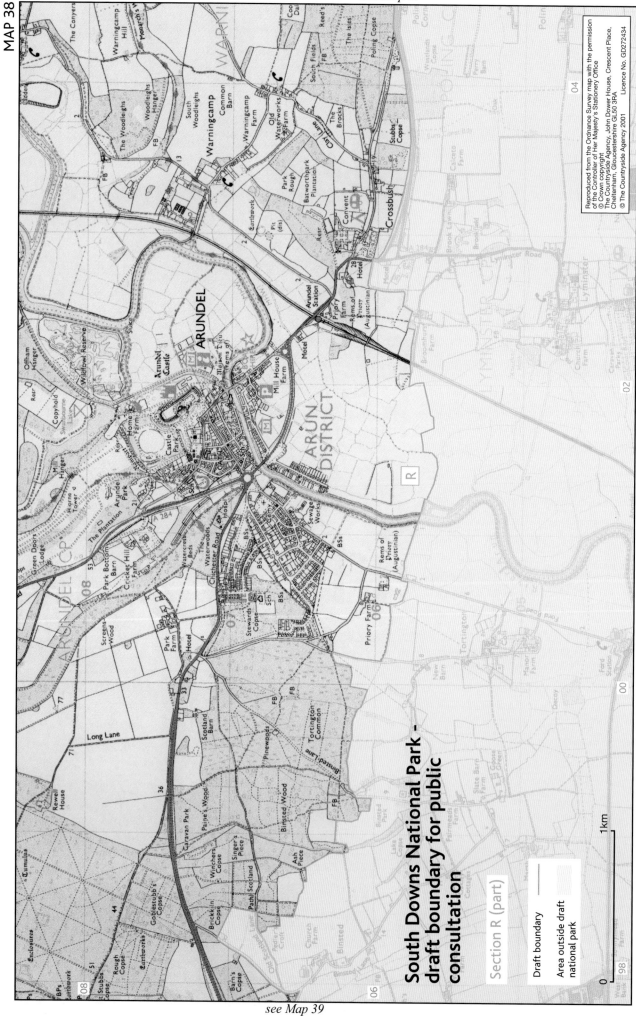

South Downs National Park – draft boundary for public consultation

Section R (part)

Draft boundary

Area outside draft national park

0   1km

Reproduced from the Ordnance Survey map with the permission of the Controller of Her Majesty's Stationery Office
© Crown copyright
The Countryside Agency, John Dower House, Crescent Place, Cheltenham, Gloucestershire GL50 3RA
© The Countryside Agency 2001   Licence No. GD272434

see Map 39

MAP 39

see Map 38

see Map 40

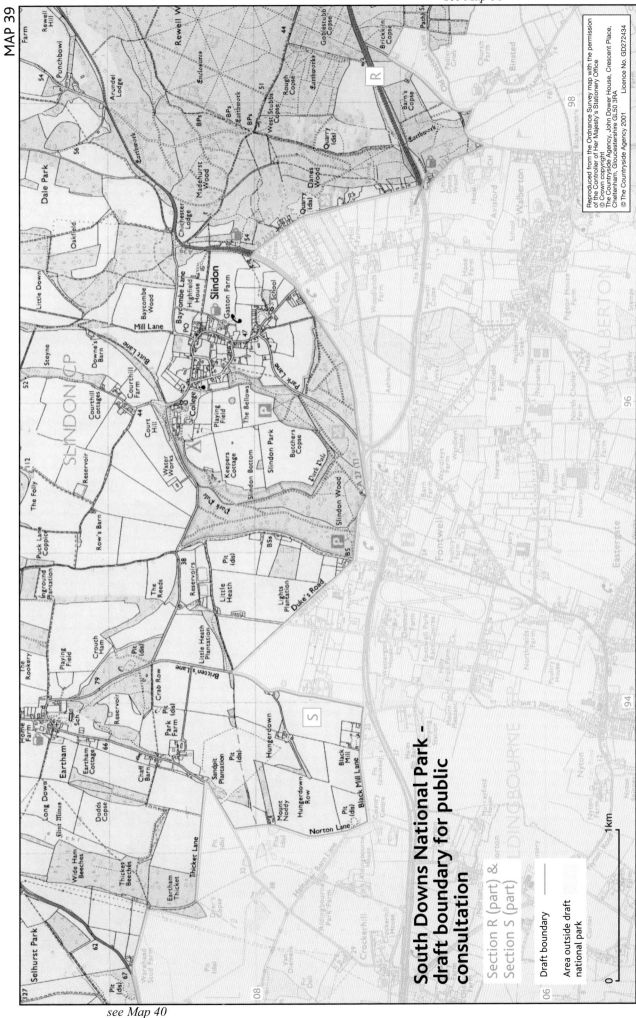

South Downs National Park - draft boundary for public consultation

Section R (part) &
Section S (part)

—————— Draft boundary

Area outside draft
national park

0   1km

see Map 39

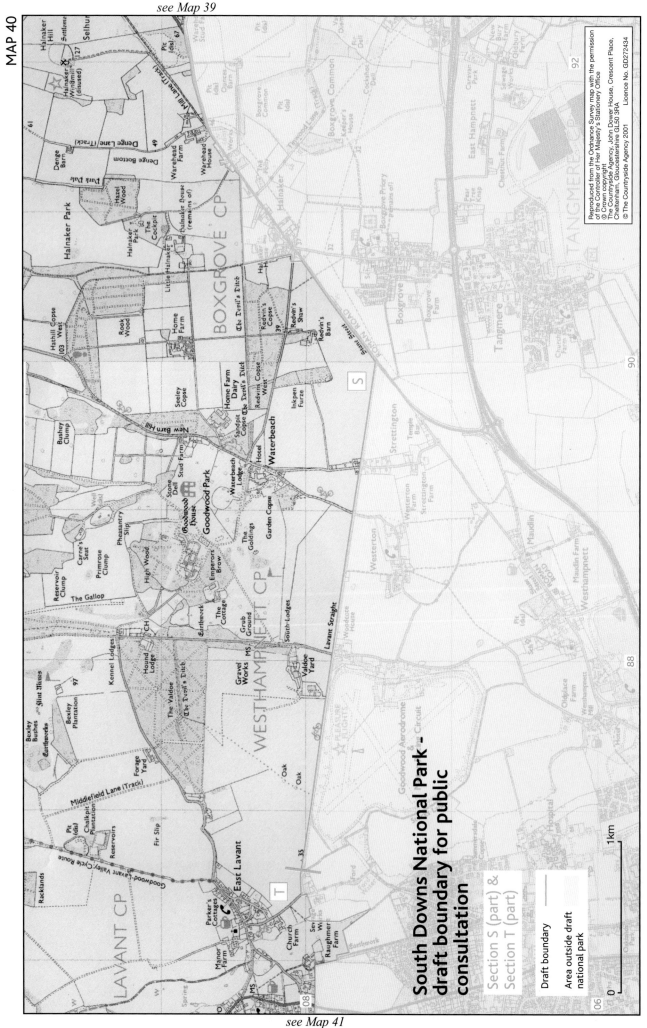

MAP 40

South Downs National Park – draft boundary for public consultation

Section S (part) & Section T (part)

Draft boundary

Area outside draft national park

0   1km

see Map 41

MAP 41

see Map 40

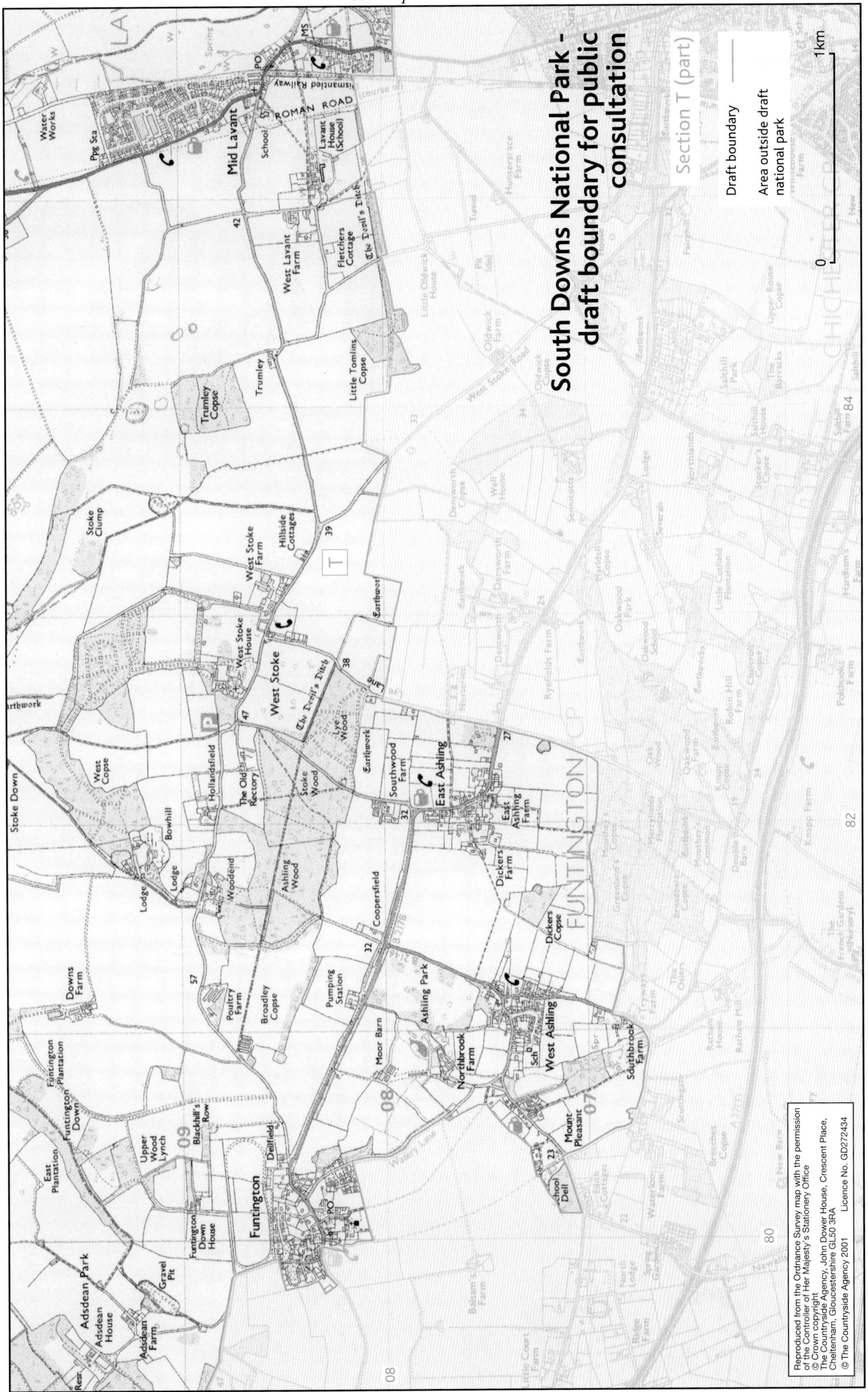

South Downs National Park - draft boundary for public consultation

Section T (part)

Draft boundary

Area outside draft
national park

0   1 km

Water Works

Pig Sta

ROMAN ROAD

Mid Lavant

School

Lavant House School

West Lavant Farm

Fletchers Cottage

The Devil's Ditch

42

Trumley Copse

Trumley

Little Tomlins Copse

Little Oldwick House

Oldwick Farm

Pit (dis)

West Stoke Road

Humbersace Farm

Tarell

Stoke Clump

Stoke Down

West Copse

West Stoke Farm

Hillside Cottages

West Stoke House

West Stoke

The Devil's Ditch

Lane

Earthwork

39

T

38

Lye Wood

Earthwork

Southwood Farm

East Ashling

Earthwork

32

East Ashling Farm

Denworth Copse

Denworth Farm

Well House

Lodge

Heldfield Copse

Oakwood Park

Oakwood School

Little Cotland Plantation

FUNTINGTON CP

Dickers Farm

Dickers Copse

27

Ryefields Farm

Nurcombe

Stecker's Copse

Salthill House

Salthill Park

The Barracks

Salthill Farm 84

Robin Hill

Knapp Copse

Crptcroft Copse

Polehooks Farm

Knapp Farm

82

CHICHESTER

Stoke Down

Lodge

Lodge

Bowhill

Hollandsfield

The Old Rectory

Stoke Wood

47

Woodend

Ashling Wood

Coopersfield

32

Pumping Station

Moor Barn

Ashling Park

Northbrook Farm

West Ashling

Sch

D

Southbrook Farm

Mount Pleasant

07

80

Downs Farm

57

Poultry Farm

Broadley Copse

Water Lane

23

School Dell

Bath Cottage

Funtington Plantation

East Plantation

Adsdean Park

Adsdean House

Adsdean Farm

Gravel Pit

Funtington Down

Funtington Plantation

Upper Wood Lynch

Blackmill's Row

09

Dellfield

Funtington

Funtington Down House

PO

08

North Lodge

Balsam's Farm

Little Court Farm

Ratham House

Ratham Mill

80

08

09

08

Reproduced from the Ordnance Survey map with the permission
of the Controller of Her Majesty's Stationery Office
© Crown copyright
The Countryside Agency, John Dower House, Crescent Place,
Cheltenham, Gloucestershire GL50 3RA
© The Countryside Agency 2001   Licence No. GD272434

see Map 42

MAP 42

see Map 42

see Map 43

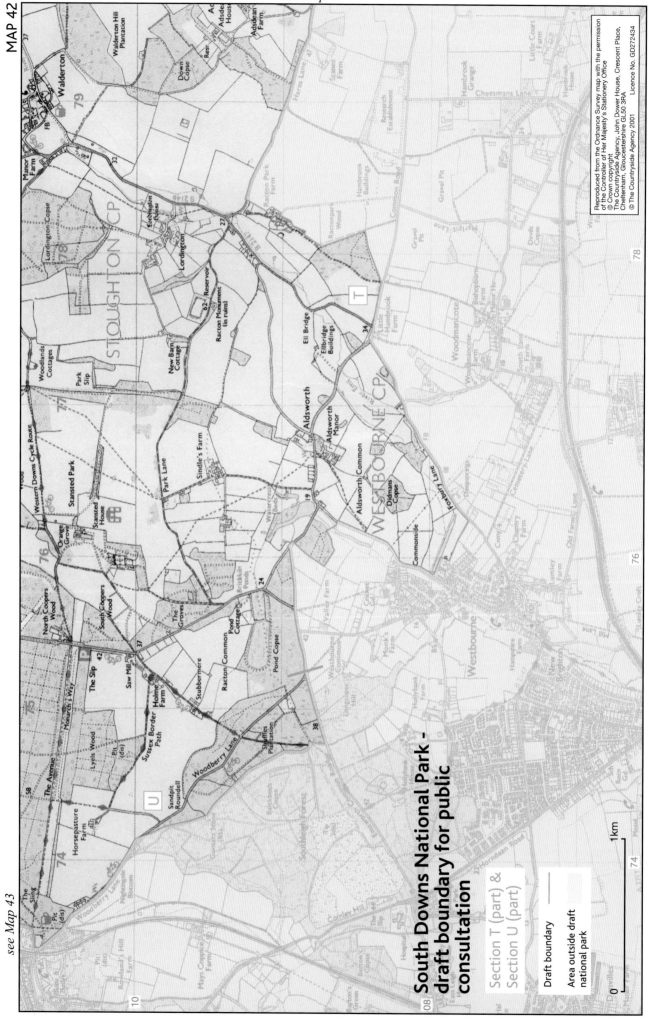

South Downs National Park – draft boundary for public consultation

Section T (part) & Section U (part)

Draft boundary

Area outside draft national park

1km

0   5miles

MAP 43

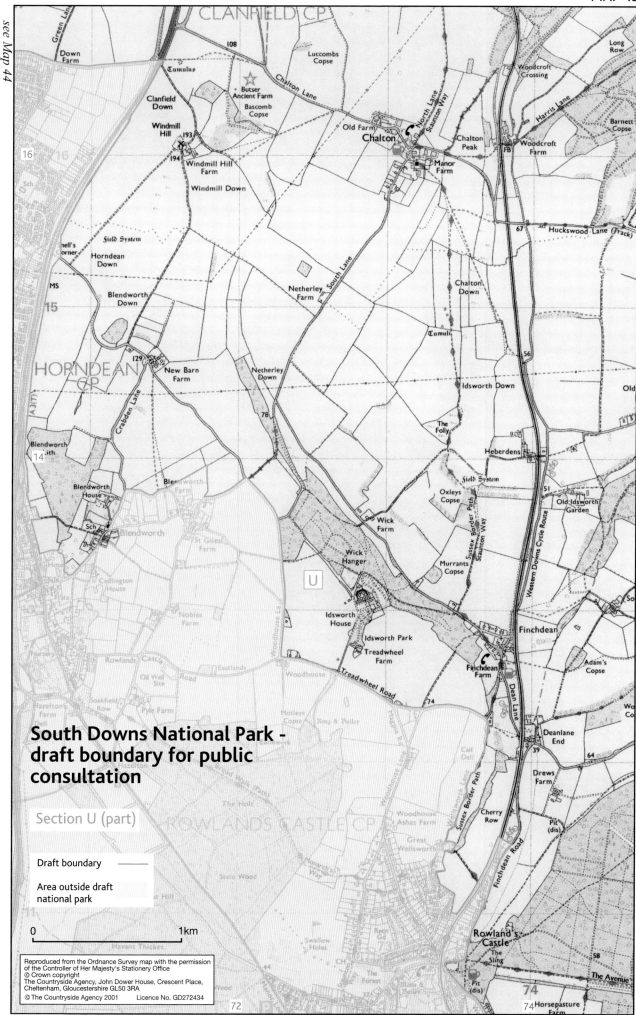

see Map 44

South Downs National Park - draft boundary for public consultation

Section U (part)

Draft boundary —————

Area outside draft national park

0 ———————— 1km

see Map 42

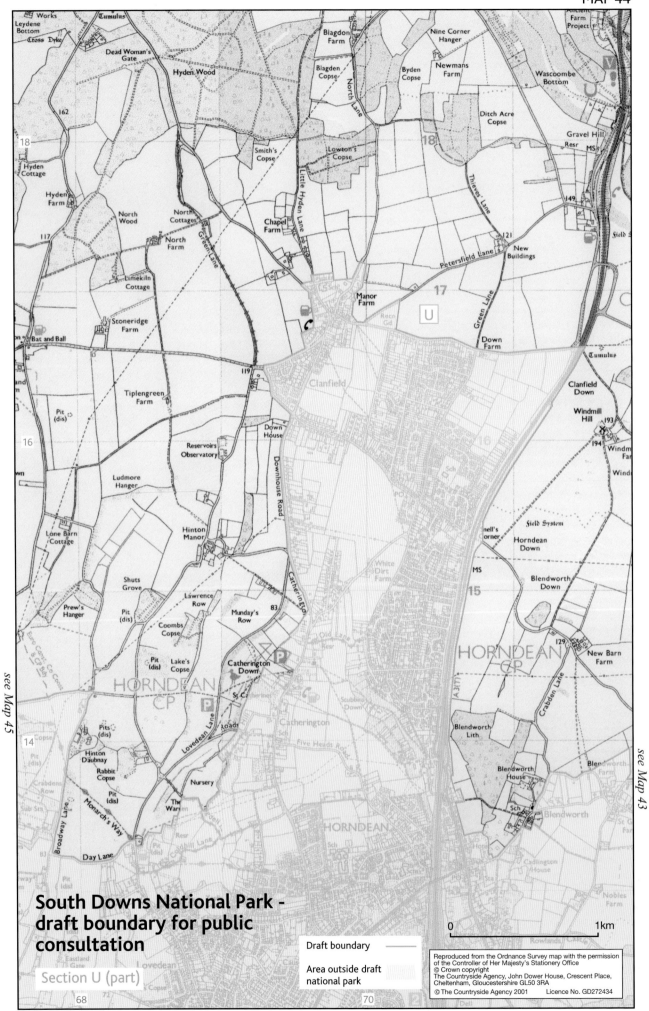

South Downs National Park - draft boundary for public consultation

Section U (part)

Draft boundary ————

Area outside draft national park

0   1km

Reproduced from the Ordnance Survey map with the permission of the Controller of Her Majesty's Stationery Office
© Crown copyright
The Countryside Agency, John Dower House, Crescent Place, Cheltenham, Gloucestershire GL50 3RA
© The Countryside Agency 2001   Licence No. GD272434

see Map 45

see Map 43

MAP 45

see Map 44

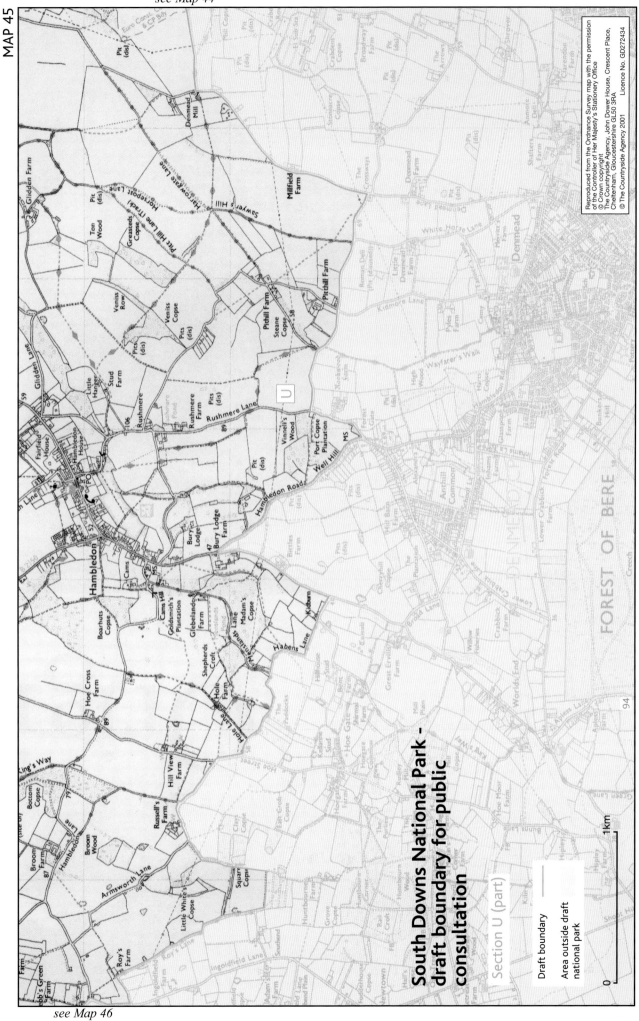

South Downs National Park – draft boundary for public consultation

Section U (part)

Draft boundary

Area outside draft national park

0   1km

see Map 46

South Downs National Park - draft boundary for public consultation

Section U (part), Section V & Section W (part)

Draft boundary ——————

Area outside draft national park

see Map 45

Swanmore

Hill Place

Hill Farm Orchards

Bottom Copse

Soberton

Cottages

Chalk Hill

Chalk Pit

Hill Grove

W

Droxford Road

King's Way

Green Lane

Wickham Road

West St

Meon Valley

Cole Hill

16

The Bungalow

SWANMORE CP

Tudor Cottage

SOBER

16

Hillpound

Cott Street

Thatched Cottages

Cott Street Farm

Cott Street Lane

St Clair's Farm

Peststead Lane Pe

Broadlands Farm

Hurricane Farm

Hamblebrook Farm

Dirty Copse

Holywell Road

Holywell House

Ragnals Copse

39

Webb's Green Farm

Longridge Farm

Forest Lodge

Little Bere

Dean Row

Roy's Farm

U

Bishopsmore

57

Bishop's Inclosure

Bishopwood Farm

Timber Yard

Holy Well

41

15

Recn Gd

78

Hawk's Nest Farm

Holy Well

Soberton Mill

FB

71a

Soberton Heath

Ingoldfield Farm

Dradfield Copse

Adam's Farm

Daysh's Farm

Mislingford

Timber Yard

Bere Farm

Dradfield Lane

Southend Plain

14

Newmans Lane

Bishop's Wood

Upperford Copse

Woodend

Liberty Road

PC

Dradfield Copse

Kingsmead

PC

X P

Frith Lane End

Kiln Copse

Newtown

Frith Lane

The Roebuck Inn

West Walk

13

Clamp Farm

Lodge Hill

Lodge Hill

Hall's Copse

Fodderhouse Copse

Close Wood

Oak Tree Farm

Frith Farm

Pit

Lodge Hill

West Lodge

97

Clamp Kiln

Charles Wood

Retreat Farm

Northfields Farm

West Lodge

Cutlers

X P

PC

42

Goathouse Farm

12

Caravan Site

Ivy Cottage

Oak

FB

Rookesbury Park Farm

School

78

Hundred Acres

P

Rookesbury Park

Wickham

V

Winscombe

North Boarhunt

Little Forest

Quob

Castle Farm

58

Mellishes Bottom

Wickham Common

North Boarhunt

Pounds

Hale Cross

0   1km

ROMAN ROAD (course of)

Birching Copse

Wine Cross

Orchard Copse

60

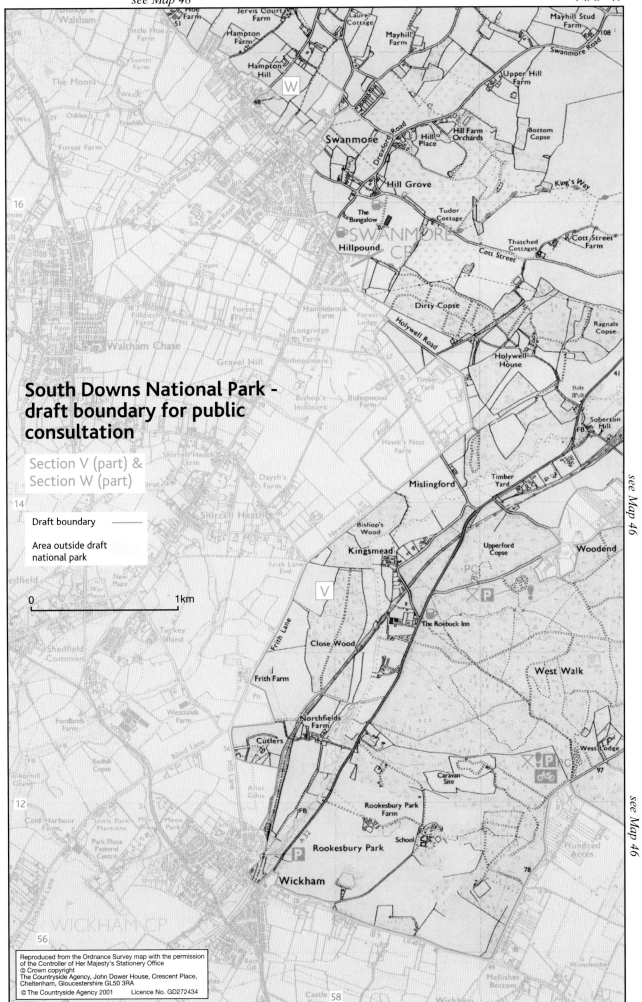

South Downs National Park -
draft boundary for public consultation

Section V (part) &
Section W (part)

Draft boundary ————

Area outside draft
national park

0   1km

see Map 46

see Map 46

MAP 48

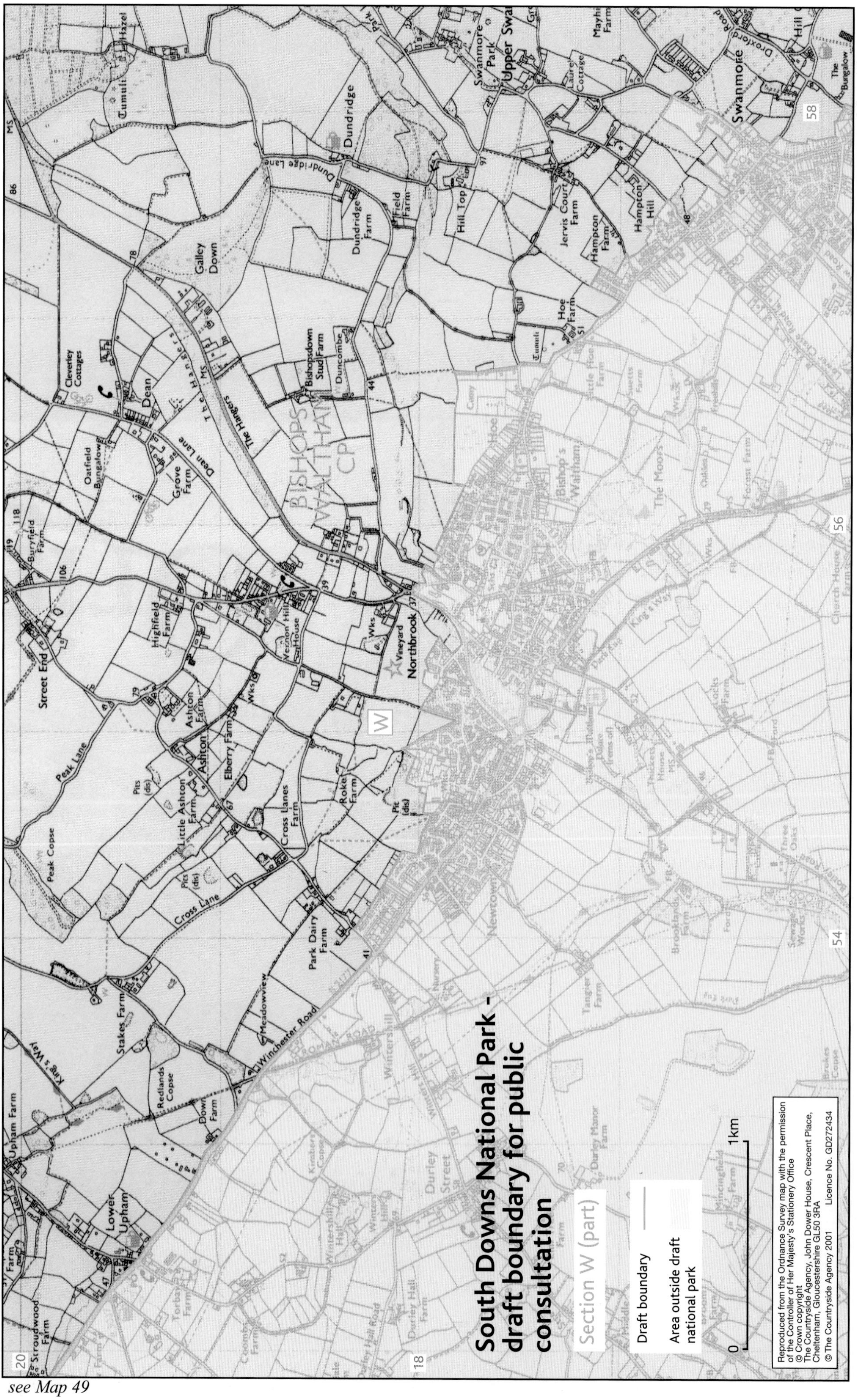

South Downs National Park – draft boundary for public consultation

Section W (part)

Draft boundary

Area outside draft national park

0   1km

see Map 47

see Map 49

MAP 49

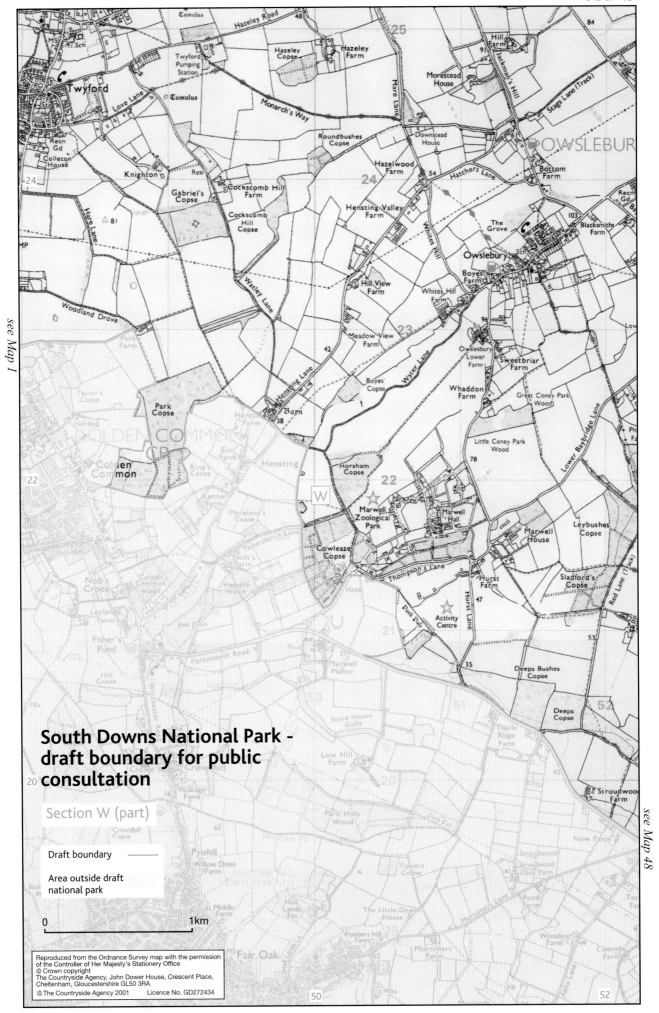

South Downs National Park - draft boundary for public consultation

Section W (part)

Draft boundary

Area outside draft national park

0 _____ 1km

see Map 1

see Map 48